Functional Programming in Go

Apply functional techniques in Golang to improve the testability, readability, and security of your code

Dylan Meeus

BIRMINGHAM—MUMBAI

Functional Programming in Go

Group Product Manager: Gebin George
Publishing Product Manager: Pooja Yadav
Senior Editor: Ruvika Rao
Technical Editor: Maran Fernandes
Copy Editor: Safis Editing
Project Coordinator: Deeksha Thakkar
Proofreader: Safis Editing
Indexer: Pratik Shirodkar
Production Designer: Joshua Misquitta
DevRel Marketing Coordinator: Sonia Chauhan

First published: March 2023
Production reference: 1270223

Published by Packt Publishing Ltd.
Livery Place
35 Livery Street
Birmingham
B3 2PB, UK.

ISBN 978-1-80181-116-3

www.packtpub.com

Contributors

About the author

Dylan Meeus is a software engineer, with over a decade of experience building software using various functional and object-oriented programming languages. He has used Go to develop systems in a variety of domains, from healthcare to machine learning and digital signal processing software. He developed a passion for functional programming when learning Haskell and applied this knowledge to traditionally non-functional languages, such as Java. Over the past several years, Dylan has been a speaker at various Go and Java-oriented conferences such as GopherCon and Devoxx.

First and foremost, I would like to thank my wife, Ana, for her words of encouragement and support throughout the process of writing this book, as well as our dog, Bucky, for letting me know when it was time to take a break and go for a walk. I also want to thank Tom and the team at Packt, whose technical advice and reviews helped shape this book.

About the reviewer

Tom Deboosere is a software developer with over 10 years of experience in the healthcare sector. He started out as a C++ developer, pivoting to Java and then Go.

He currently works for nexuzhealth as a full-time technical domain lead, working on software used by 35+ care facilities, with over a million users.

He also likes to cook, play board games, and take long walks on the beach...

...as long as the beach is in virtual reality.

Table of Contents

5

Immutability   79

Part 2: Using Functional Programming Techniques

6

Three Common Categories of Functions   99

7

Recursion   119

8

Readable Function Composition with Fluent Programming   139

Part 3: Design Patterns and Functional Programming Libraries

9

Functional Design Patterns   163

Preface

Go is a multi-paradigm programming language. This means that both the object-oriented paradigm and the functional paradigm are entirely valid approaches to problem solving. In this book, we will explore the applications of functional programming techniques in Go. But rather than being purely focused on the functional aspect, we will embrace Go for what it is – multi-paradigm. This means that we highlight the difference between the functional and object-oriented ways of problem solving.

To write Go code that is more testable, readable, and reliable, we will look at functional-first approaches such as functions as first-class citizens, function purity, currying, and more. We will look not only at how to write functional code, but we will also explore the performance implications and limitations of Go.

The goal of this book is to get the reader accustomed to functional programming as a valid paradigm that can improve your code, no matter whether you're working on a greenfield project or a project already entrenched in the OO paradigm.

For readers unfamiliar with the newly introduced generics in Go, this book also serves as an example of what's possible now that generics are part of the standard library. Finally, we will also look at libraries that can be leveraged to write functional code for both pre-generic and post-generic versions of Go.

Who this book is for

If you are a Go engineer with a background in traditionally object-oriented languages such as Java or C++ who wants to broaden your knowledge of functional programming, this book is for you. The book aims to teach you how concepts from functional programming can improve your existing Go code, as well as when to choose the functional approach. At each step, we highlight the trade-offs between the functional and object-oriented approaches to see how they compare.

What this book covers

In *Chapter 1, Introducing Functional Programming*, we are going to take a bird's eye view of the *what* and *why* behind functional programming. To start, we will take a brief look at the history and contemporary state of functional programming methodologies. Then we will take a look at how functional programming compares to the more traditional object-oriented programming.

In *Chapter 2, Treating Functions as First-Class Citizens*, we are going to cover exactly why functions are powerful in languages that treat them as **first-class citizens**. Go has functions as first-class citizens out of the box, meaning we get this functionality. We are going to see how this allows us to create function-centered constructs that improve the readability and testability of our code.

In *Chapter 3, Higher-Order Functions*, we are going to explore the concept of function composition through higher-order functions. There are a variety of new concepts that are introduced here, such as closures, partial application, and function currying. We will take a look at some practical examples and real-world use cases for these.

In *Chapter 4, Write Testable Code with Pure Functions*, we will take a look at what it means for a language, and a function, to be considered pure. We will take a look at some of the tradeoffs between function purity and impurity, and explore how pure functions help us write testable code.

In *Chapter 5, Immutability*, we cover what exactly it means to be immutable, and how the Go language can help preserve immutability at the struct level. To understand how this works, we will look at how Go handles pointers and references to objects, what the performance implications are, and how to decide between the pointer-reference trade-offs. We will also dive into the implications for garbage collection, unit testing, and pure functional programming.

In *Chapter 6, Three Common Categories of Functions*, we are going to look at some practical implementations of functions that leverage the concepts of functional programming covered up to now. We will build Filter functions, Map functions and Reducers.

In *Chapter 7, Recursion*, we are going to talk about recursion. This is a topic that all programmers encounter sooner or later, as it's not exclusive to the functional paradigm. Any language in which you can express function calls also allows you to express functions that are recursive in nature. But in functional languages, these take center stage. We will look at the implications for this in Go.

In *Chapter 8, Readable Function Composition with Fluent Programming*, we are going to look at different methods for chaining functions in functional programming. The end goal here is to write code that is easier to read and creates less visual clutter. We will look at three ways for achieving this. First, we will take a look at how we can use type aliases to attach methods to container types, allowing us to create chained functions with the familiar *dot notation*. Next, we will look at continuation-passing style programming and consider the trade-offs of each approach.

In *Chapter 9, Functional Design Patterns*, we will move to a higher level of abstraction. Rather than talking about individual functions and operations, we will look at design patterns. While we will not extensively explain each design pattern, we will take a look at how the object-oriented pattern translates to the functional world.

In *Chapter 10, Concurrency and Functional Programming*, we consider how concurrency is all around us, both in the real world as well as the virtual one. In this chapter we will start by looking at concurrency, parallelism, and distributed computation. Next, we will focus on how the concurrency mechanisms in Go can help us write functional code.

In *Chapter 11, Functional Programming Libraries*, we will explore several libraries that can help us build programs in the functional paradigm. We will look both at pre-generic libraries and post-generic libraries.

To get the most out of this book

Prior to picking up this book, the reader should be familiar with Go and generics. The basic concepts of the programming language (control flow, structs, and imports), how to build and run applications, and how to import open source libraries from GitHub should also be understood by the reader.

Software/hardware covered in the book	Operating system requirements
Go (pre- and post-generics)	Windows, macOS, or Linux

Having Go 1.18 or later installed is a prerequisite for the majority of this book. Certain chapters will also work on Go version's prior to 1.18, this will be called out per chapter. Most of the code will also work in the Go playground at `https://go.dev/play/`.

If you are using the digital version of this book, we advise you to type the code yourself or access the code from the book's GitHub repository (a link is available in the next section). Doing so will help you avoid any potential errors related to the copying and pasting of code.

Some chapters will have snippets in Haskell and Java for illustrative purposes of (pure) functional and object-oriented counterparts to Go.

Download the example code files

You can download the example code files for this book from GitHub at `https://github.com/PacktPublishing/Functional-Programming-in-Go`. If there's an update to the code, it will be updated in the GitHub repository.

We also have other code bundles from our rich catalog of books and videos available at `https://github.com/PacktPublishing/`. Check them out!

Download the color images

We also provide a PDF file that has color images of the screenshots and diagrams used in this book. You can download it here: `https://packt.link/5tPDg`.

Conventions used

There are a number of text conventions used throughout this book.

`Code in text`: Indicates code words in text, database table names, folder names, filenames, file extensions, pathnames, dummy URLs, user input, and Twitter handles. Here is an example: "When calling the `rollDice` function, the output is not consistent. If it were consistently outputting the same number, it would be a pretty bad randomization function."

A block of code is set as follows:

```
func rollDice() int {
    return rand.Intn(6)
}
```

Any command-line input or output is written as follows:

```
go test -bench=.
```

Bold: Indicates a new term, an important word, or words that you see onscreen. For instance, words in menus or dialog boxes appear in **bold**. Here is an example: "In this main function, we are first defining a **deferred** function that runs at the end of main function, just before function exit."

> **Tips or important notes**
> Appear like this.

Get in touch

Feedback from our readers is always welcome.

General feedback: If you have questions about any aspect of this book, email us at customercare@packtpub.com and mention the book title in the subject of your message.

Errata: Although we have taken every care to ensure the accuracy of our content, mistakes do happen. If you have found a mistake in this book, we would be grateful if you would report this to us. Please visit www.packtpub.com/support/errata and fill in the form.

Piracy: If you come across any illegal copies of our works in any form on the internet, we would be grateful if you would provide us with the location address or website name. Please contact us at copyright@packt.com with a link to the material.

If you are interested in becoming an author: If there is a topic that you have expertise in and you are interested in either writing or contributing to a book, please visit authors.packtpub.com.

Share your thoughts

Once you've read *Functional Programming in Go*, we'd love to hear your thoughts! Scan the QR code below to go straight to the Amazon review page for this book and share your feedback.

https://packt.link/r/9781803238012

Your review is important to us and the tech community and will help us make sure we're delivering excellent quality content.

Download a free PDF copy of this book

Thanks for purchasing this book!

Do you like to read on the go but are unable to carry your print books everywhere?

Is your eBook purchase not compatible with the device of your choice?

Don't worry, now with every Packt book you get a DRM-free PDF version of that book at no cost.

Read anywhere, any place, on any device. Search, copy, and paste code from your favorite technical books directly into your application.

The perks don't stop there, you can get exclusive access to discounts, newsletters, and great free content in your inbox daily

Follow these simple steps to get the benefits:

1. Scan the QR code or visit the link below

https://packt.link/free-ebook/9781801811163

2. Submit your proof of purchase
3. That's it! We'll send your free PDF and other benefits to your email directly

Part 1: Functional Programming Paradigm Essentials

In this part, we will take a look at what the functional programming paradigm entails. We'll look at how it compares to the traditional object-oriented approach, and learn some language design differences between programming languages in each paradigm. We'll also discuss what it means for Go to be a multi-paradigm language and see how this benefits our use case. Finally, we'll look at some key ideas in functional programming, which we can leverage to write more readable, maintainable, and testable code.

This part has the following chapters:

1

Introducing Functional Programming

In this first chapter, we are going to take a bird's eye view of the *what* and *why* behind **functional programming** (**FP**). Before we dive into the nitty gritty of FP, we first have to understand what benefit we get from applying these techniques to our code. To start off, we will provide a brief look into the history and contemporary state of FP methodologies. Next, we will take a look at how FP compares to more traditional **object-oriented programming** (**OOP**). Finally, we will also discuss the "Go programming paradigm."

The main things we will cover in this chapter are as follows:

- What is FP?
- A brief history of FP
- A look at the current state of FP
- A comparison of traditional object-oriented and functional methodologies
- A discussion on Go programming paradigms and how FP fits into this

What is functional programming?

As you might have guessed, FP is a programming paradigm where functions play the main role. Functions will be the bread and butter of the functional programmer's toolbox. Our programs will be composed of functions, chained together in various ways to perform ever more complex tasks. These functions tend to be small and modular.

This is in contrast with OOP, where objects play the main role. Functions are also used in OOP, but their use is usually to change the state of an object. They are typically tied to an object as well. This gives the familiar call pattern of *someObject.doSomething()*. Functions in these languages are treated as secondary citizens; they are used to serve an object's functionality rather than being used for the function itself.

Introducing first-class functions

In FP, functions are considered **first-class citizens**. This means they are treated in a similar way to how objects are treated in a traditional object-oriented language. Functions can be bound to variable names, they can be passed to other functions, or even served as the return value of a function. In essence, functions are treated as any other "type" would be. This equivalence between types and functions is where the power of FP stems from. As we will see in later chapters, treating functions as first-class citizens opens a wide door of possibilities for how to structure programs.

Let's take a look at an example of treating functions as first-class citizens. Don't worry if what's happening here is not entirely clear yet; we'll have a full chapter dedicated to this later in the book:

```go
package main
import "fmt"
type predicate func(int) bool
func main() {
    is := []int{1, 1, 2, 3, 5, 8, 13}
    larger := filter(is, largerThan5)
    fmt.Printf("%v", larger)
}
func filter(is []int, condition predicate) []int {
    out := []int{}
    for _, i := range is {
        if condition(i) {
            out = append(out, i)
        }
    }
    return out
}
func largerThan5(i int) bool {
    return i > 5
}
```

Let's break what's happening here down a bit. First, we are using a "type alias" to define a new type. The new type is actually a "function" and not a primitive or a struct:

```go
type predicate func(int) bool
```

This tells us that everywhere in our code base where we find the `predicate` type, it expects to see a function that takes an `int` and returns a `bool`. In our `filter` function, we are using this to say we expect a slice of integers as input, as well as a function that matches the `predicate` type:

```
func filter(is []int, condition predicate) []int {…}
```

These are two examples of how functions are treated differently in functional languages from in object-oriented languages. First, types can be defined as functions instead of just classes or primitives. Second, we can pass any function that satisfies our type signature to the `filter` function.

In the `main` function, we are showing an example of passing the `isLargerThan5` function to the `filter` function, similar to how you'd pass around objects in an object-oriented language:

```
larger := filter(is, largerThan5)
```

This is a small example of what we can do with FP. This basic idea, of treating functions as just another type in our system that can be used in the same way as a struct, will lead to the powerful techniques that we explore in this book.

What are pure functions?

FP is often thought of as a purely academic paradigm, with little to no application in industry. This, I think, stems from an idea that FP is somehow more complicated and less mature for the industry than OOP. While the roots of FP are academic, the concepts that are central to these languages can be applied to many problems that we solve in industry.

Often, FP is thought of as more complex than traditional OOP. I believe this is a misconception. Often, when people say FP, what they really mean to say is *pure FP*. A pure functional program is a subset of FP, where each function has to be pure – it cannot mutate the state of a system or produce any side effects. Hence, a pure function is completely predictable. Given the same set of inputs, it will always produce the same set of outputs. Our program becomes entirely deterministic.

This book will focus on FP without treating it as the stricter subset of "pure" FP. That is not to say that purity brings us no value. In a purely functional language, functions are entirely deterministic and the state of a system is unchanged by calling them. This makes code easier to debug and comprehend and improves testability. *Chapter 6* is dedicated to function purity, as it can bring immense value to our programs. However, eradicating all side effects from our code base is often more trouble than it's worth. The goal of this book is to help you write code in a way that improves readability, and as such, we'll often have to make a trade-off between the (pure) functional style and a more forgiving style of FP.

To briefly and rather abstractly show what function purity is, consider the following example. Say we have a struct of the `Person` type, with a `Name` field. We can create a function to change the name of the person, such as `changeName`. There are two ways to implement this:

- We can create a function that takes in the object, changes the content of the `name` field to the new name, and returns nothing.
- We can create a function that takes in an object and returns a new object with the changes applied. The original object is not changed.

The first way does not create a pure function, as it has changed the state of our system. If we want to avoid this, we can instead create a `changeName` function that returns a new `Person` object that has identical field values for each field as the original `Person` object does, but instead has a new name in the `name` field. The diagram here shows this a bit more abstractly:

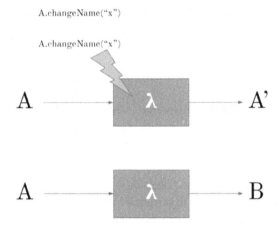

Figure 1.1: Pure function (top) compared to impure function (bottom)

In the top diagram, we have a function (denoted with the Lambda symbol) that takes a certain object, **A**, as input. It performs an operation on this object, but instead of changing the object, it returns a new object, **B**, which has the transformation applied to it. The bottom diagram shows what was explained in the earlier paragraph. The function takes object **A**, makes a change "in-place" on the object's values, and returns nothing. It has only changed the state of the system.

Let's take a look at what this would look like in code. We start off by defining our struct, `Person`:

```
type Person struct {
    Age    int
    Name   string
}
```

To implement the function that mutates the `Person` object and places a new value in the `Name` field, we can write the following:

```
func changeName(p *Person, newName string) {
    p.Name = newName
}
```

This is equivalent to the bottom of the diagram; the `Person` object that was passed to the function is mutated. The state of our system is now different from before the function was called. Every place that refers to that `Person` object will now see the new name instead of the old name.

If we were to write this in a pure function, we'd get the following:

```
func changeNamePure(p Person, newName string) Person {
    return Person{
        Age:  p.Age,
        Name: newName,
    }
}
```

In this second function, we copy over the `Age` value from the original `Person` object (p) and place the `newName` value in the `Name` field. The result of this is returned as a new object.

While it's true that the former, impure way of writing code seems easier superficially and takes less effort, the implications for maintaining a system where functions can change the state of the system are vast. In larger applications, maintaining a clear understanding of the state of your system will help you debug and replicate errors more easily.

This example looks at pure functions in the context of immutable data structures. A pure function will not mutate the state of our system and always return the same output given the same input.

In this book, we will focus on the essence of FP and how we can apply the techniques in Go to create more readable, maintainable, and testable code. We will look at the core building blocks, such as higher-order functions, function currying, recursion, and declarative programming. As mentioned previously, FP is not equivalent to "pure" FP, but we will discuss the purity aspect as well.

Say what you want, not how you want it

One commonality that is shared between FP languages is that functions are declarative rather than imperative. In a functional language, you, as the programmer, say *what* you want to achieve rather than *how* to achieve it. Compare these two snippets of the Go code.

The first snippet here is an example of valid Go code where the result is obtained declaratively:

```
func DeclarativeFunction() int {
    return IntRange(-10,10).
        Abs().
        Filter(func(i int64) bool {
            return i % 2 == 0
        }).
        Sum()
    // result = 60
}
```

Notice how, in this code, we say the following things:

- Give us a range of integers, between -10 and 10
- Turn these numbers into their absolute value
- Filter for all the even numbers
- Give us the sum of these even numbers

Nowhere did we say how to achieve these things. In an imperative style, the code would look like the following:

```
func iterativeFunction() int {
    sum := 0
    for i := -10; i <= 10; i++ {
        absolute := int(math.Abs(float64(i)))
        if absolute%2 == 0 {
            sum += absolute
        }
    }
    return sum
}
```

While, in this example, both snippets are easy to read for anyone with some Go experience, we can imagine how this would stop being the case for larger examples. In the imperative example, we have to spell out literally how the computer is supposed to give us a result.

A brief history of functional programming

If you take a look at the mainstream languages of the past decade, you will notice how the prevailing programming paradigm is **OOP**. This might lead you to believe that FP is an upstart paradigm, one that is in a young state compared to the well-established object-oriented approach. Yet, when we look at the history of FP, we can trace its roots all the way back to the 1930s, quite some time before we talked about programming in the modern-day sense.

The roots of FP can be traced back to the Lambda calculus, which was developed in the 1930s by Alonzo Church. This was developed as a formal system based on function abstraction and application, using variable binding. There are two variants of this calculus; it can either be **typed** or **untyped**. This is directly parallel to how programming languages today are either **statically typed**, such as Java and Go, or **dynamically typed** such as Python. The Lambda calculus was proven to be Turing-complete in 1937 – again, similar to how all mainstream programming languages today are Turing-complete.

The Lambda calculus predates modern programming by a few decades. To get to the first actual code that could be thought of as FP, in the way that we understand programming today, we have to move forward a few decades. The **LISt Processor** (**LISP**) was originally created in the 1950s as a practical application of mathematical notation. This was influenced by the Lambda calculus laid out in the 1930s by Church.

LISP can be thought of as the first FP language that reached some sense of popularity. It was especially popular in the field of artificial intelligence research, but through the decades, made its way to industry. Derivatives of LISP continued to be popular for a long time, with notable achievements such as the Crash Bandicoot game, and Hacker News being written in derivatives of this language.

LISP was developed in the late 1950s by John McCarthy. To define LISP functions, he took inspiration from the Lambda calculus developed by Church. It was extended beyond the mathematical system by introducing recursion, a fundamental concept for how functional languages work. Beyond recursion, LISP also treated functions as first-class citizens and pushed innovation in programming language design by including things such as garbage collection and conditional statements.

In the early 1960s, Kenneth E. Iverson developed **A Programming Language** (**APL**). APL is, again, a functional language that is perhaps most known for its use of symbols and terse code. For example, the following is an image of the code snippet that would generate Conway's Game Of Life:

Figure 1.2: Conway's Game of Life in APL

Jumping ahead another decade, in 1973, we get a language called **Meta Language** (**ML**). This language introduced the *polymorphic Hindley-Milner type system* – that is to say, a type system in which types are assigned automatically without requiring explicit type annotations. In addition, it supported features such as **function currying**, which we will apply to our functional Go code later in this book. It also supports pattern matching on the arguments of a function, as we can see in the following snippet of a function to compute the factorial of a number:

```
fun fac 0 = 1
  | fac n = n * fac (n - 1)
```

The pattern matcher in this example will take a look at what the input value is to the `fac` function, and then either continue with the first line if the input value is 0, or the second line in all other cases. Notice that this is also a recursive function expressed quite beautifully. Sadly, pattern matching will not be explored further in this book, as Go currently offers no way of doing this. We will see a way of doing a similar type of function dispatching using maps and higher-order functions.

In 1977, the language called FP was created. It was developed by John Backus specifically to support the FP paradigm. While the language itself did not get much traction outside of academia, the paper in which it was introduced (*Can programming be liberated from the von Neumann style?*) did spark a renewed interest in FP.

In the same decade as ML and FP, another language called **Scheme** was developed. This is the first dialect of LISP that used lexical scoping and tail-call optimization. Tail-call optimization led to the practical implementation of recursive algorithms. While the details of tail-call optimization will be discussed in *Chapter 7* of this book, briefly stated, it allows recursive algorithms to be implemented in an efficient way and without using more memory than a traditional loop would, thus eliminating the "stack overflow exceptions" that otherwise would happen during deep recursion.

Scheme is one of the most influential LISP dialects to have been created, and continues to this day to have some popularity. Although it was created in 1975, the last standard was defined as recently as 2013 (R7RS-Small). Scheme, in turn, influenced other LISP dialects, the most notable of which is perhaps Common Lisp. Interestingly, although having roots in FP, Common Lisp introduced the **Common Lisp Object System (CLOS)**. The CLOS facilitated OOP in LISP. With this, we can perhaps consider LISP a truly multi-paradigm language, not unlike Go.

The final language to look at before we jump to contemporary functional languages is Miranda. Miranda is a lazy, purely FP language. The key concept that was introduced here is lazy evaluation. When a language is said to support lazy evaluation, this means that an expression is not resolved until the value is actually needed. It can be used, for example, to implement infinite data structures. You could define a function that generates all Fibonacci numbers, a sequence that never ends – but, rather than creating the entire list (which is not possible), it will only generate a subset of that list that is relevant to the problem you are solving. For example, the following snippet of Miranda code computes all square numbers:

```
squares = [ n*n | n <- [0..] ]
```

With that, we have arrived at the next language to discuss briefly, namely Haskell.

Modern functional programming

After our brief look at the history of FP, it's time to dive into modern functional languages. One of the languages that is popular today within the strict FP languages is Haskell. When people study FP or become exposed to it, it is often through this language. Haskell is a statically typed FP language. It has goodies such as type inference (like ML) and lazy evaluation (like Miranda).

When people want to learn more about pure FP, my recommendation is always to start with Haskell. It has a great community and plenty of resources and teaches you all there is to know about the FP domain.

It might very well be the most popular pure FP language around, yet it accounts for less than 1% of the active users on GitHub (`https://www.benfrederickson.com/ranking-programming-languages-by-github-users/`). For fun, if we take a look at Go, it currently sits at about approximately 4% of active users. Not bad for a language that's just about a decade old at this point!

In the .NET world, another language that is relatively popular is F#. While this is not a purely functional language such as Haskell, it is a *functional-first* language. It prefers the functional paradigm over others but does not enforce it. Similarly to Haskell, it has less than 1% of active users on GitHub. C# seems to get all the popular features of F# though, so at least the functional concepts that F# spearheaded for .NET will find popularity.

So does that mean functional programming is dead on arrival? Well, not quite. The book you are reading now is all about Go, and Go is not a purely FP language. My take on it is that the concepts from FP are generally useful and can create better code, even in object-oriented languages – and I'd like to think I'm not alone in thinking this. Many of the languages that we think of as object-oriented languages have become more and more functional.

We can see this shift happening even in the most popular mainstream object-oriented languages. Java is introducing FP concepts with each iteration, offering such things as pattern matching, higher-order functions, and declarative programming through Lambda functions. C# is looking more and more like F# (Microsoft's functional programming counterpart of C#) with each release. They have implemented pattern matching, immutability, built-in tuple support, and more.

This shift is happening because, although purely functional languages might not always suit the industry, the concepts from functional languages allow us to write our object-oriented code with more confidence. They lead to code that is easier to test, easier to read, and faster to debug.

The most popular programming language used today is JavaScript. While this would perhaps not pop into people's minds when talking about FP, it does meet a subset of the "requirements" we have for a functional language. It has the following:

- First-class functions
- Anonymous (Lambda) functions
- Closures

When combining these features, we can create many constructs that allow us to leverage code in an FP style.

For those of us who want to have a purely functional language in the browser, there are languages that transpile to JavaScript, such as Elm and PureScript.

Let's now take a look at the star of this book, Go, and how this fits into the picture.

The Go programming paradigm

Unless this is your first introduction to Go, you probably know that Go is a statically typed programming language. You also know that it has structs and that we can instantiate objects out of these. You likely also know that Go optionally binds functions to a struct, but that is not required. It would be possible to write an entire Go program without creating an object, something that the stricter object-oriented languages rarely allow.

In fact, the simplest `Hello World` program in Go has no sense of structs or objects:

```
package main
import "fmt"
func main() {
    fmt.Println("Hello Reader!")
}
```

As you can see, the introductory Go program that many of us wrote when starting to learn Go has no notion of structs or objects to do something useful. `Println` is a function defined in the `fmt` package, but it's not bound to an object.

The term for a language such as Go is multi-paradigm. Go does not force us to write code in the object-oriented paradigm or in the functional paradigm. We, the programmers, have complete freedom to use the language however we want. This is why the book you are reading right now exists.

Go offers several features that enable us to write functional Go code with (relative) ease:

- Functions as first-class citizens
- Higher-order functions
- Immutability guarantees
- Generics (not needed per se, but make life easier)
- Recursion

These are explored in more detail later in the book. I also want to point out some features that Go lacks (as of 1.18) that would improve our quality of life:

- Tail-call optimization

- Lazy evaluation

- Purity guarantee

These are not deal-breakers. The focus of this book is leveraging FP in Go to write better code. Even if we don't have a purely statically typed system to work with, we can work with what we do have.

By no means do I want to posit FP as the superior way to write Go code. Nor do I want to frame it as the "right" paradigm to choose. Go is multi-paradigm, and just as programmers choose the right language for any problem, we also have to choose the right paradigm for each problem. We can even opt to stick to functional concepts 90% of the time and still end up with cleaner code than if we had stuck to it 100%. For example, writing purely functional code would prevent the use of any side effects. Yet, many side effects do serve a purpose. Any time we want to show a user output or get input from a user, we are technically dealing with a side effect.

Why functional programming?

All this does not yet tell us why we want to invest time in learning about FP. The main benefits we hope to get from functional programming are as follows:

- More readable code

- Easier to understand and debug code

- Easier testing

- Fewer bugs

- Easier concurrency

These can be achieved by a relatively small set of FP features. To achieve more readable code, this can be done by writing code in a declarative way. Declarative programming will show us what is happening rather than how it is happening. Declarative code is often more concise than the imperative counterpart. Conciseness is not necessarily a benefit to code readability (remember the APL example previously?) but when applied correctly, it can be.

FP makes code easier to understand, debug, and test by preferring purity over impurity. When each function always creates a deterministic outcome, we can trust a function does only what it says. When you encounter a function called `square(n int)`, we can be convinced that all the function does is square the input.

In addition, the state of the system is not changed. If we are working with structs and objects, it helps us guarantee that the values our object holds are not changed by functions that are operating on it. This reduces the cognitive overhead when reasoning about our programs.

Pure, immutable code makes code easier to test for the following reasons:

- The state of the system has no impact on our function – so we don't have to mock the state when testing.
- A given function always returns the same output for a given input. This means we get predictable, deterministic functions.

I won't be advocating for test-driven development or any such thing here, but I do believe testing is critical to writing good code. Or at least, to avoid being paged at 3 A.M. because a function started throwing unintelligible error codes at a user.

Hand in hand with more testable code, FP helps us write fewer bugs. This is perhaps hard to quantify, but the idea here is that without mutable states and with only predictable functions in our code, we'll have fewer edge cases to think about. If the state is important to your program, you have to know, at each point in time, what the state of the system can be and how it influences the function you are writing. This gets complex fast.

Finally, FP will make writing concurrent code easier. Go is a pretty well-known language for its built-in concurrency features. Concurrency was part of Go from its inception and was not tacked on later as with some other mainstream languages. As a result, Go has pretty solid concurrent coding tools.

The way in which functional programming helps is that functions are deterministic and immutable. Thus, running the same function concurrently can never impact the result of another running function. If functions never depend on the state of the system, thread A can not invalidate the system's state for thread B.

One thing I want to highlight again, as it is important, is that I don't advocate sticking to pure FP in Go. Doing so will probably make your life, and that of your coworkers, harder than it has to be. Choose the right tool for the job – sometimes that will be objects, and sometimes that will be functions.

Why not functional programming in Go?

To provide a holistic view of how FP can help us, as Go programmers, we should also consider when not to use FP. I view FP as a tool in my toolbox and when a problem lends itself to it, I will gladly use it – but just as importantly, we have to recognize when this does not work.

One of the concerns around FP is performance – while there is a lot to say on this topic, as we'll see in later chapters, performance concerns could mean we throw out some functional concepts such as immutability in favor of executing with speed. This is more complex than it might sound at first, as Go pointers are not guaranteed to be faster than Go's pass-by-value functions. We'll expand more on the performance concerns in later chapters.

Another reason not to choose FP is Go's lack of tail-call optimization. In theory, every loop you write in your program could be replaced by a recursive call, but as of Go 1.18, Go does not have the necessary tools to do this efficiently and you'd risk running into stack overflows. There are ways around this, as we will see, but if it starts sacrificing performance or readability significantly, my advice would be to just write a loop. This is not to say recursion is never the right approach. If you've worked with trees or graphs extensively, you've probably written some recursive algorithms and found them to work just fine.

Finally, if you are working on an existing code base with many other contributors, the best thing to do is follow the style of the code base. While some concepts of FP can be introduced quite easily, it is harder to enforce them in a team that's not on board with the whole idea. Luckily, many programmers today see benefits in key concepts of FP. Even in Java or C#, the idea of immutable code is embraced. Side effects similarly are more and more seen as unwanted.

Let's embrace Go as a fully multi-paradigm language and leverage each paradigm where it makes sense.

Comparing FP and OOP

As we have seen in the preceding pages, FP is not exactly a new thing. It, in fact, predates the object-oriented paradigm by a few decades. While Go is multi-paradigm and we can embrace both styles of programming, let's take a quick look at a concrete comparison between the two.

Functional programming	Object-oriented programming
Functions are the bread and butter	Classes and objects are the bread and butter
Declarative code	Imperative code
Immutability preferred	Mutable state
Can enforce purity	Often no focus on purity
Recursion	Loops

Table 1.1: Table comparing FP (left) and OOP (right)

This comparison is a tad superficial. Many object-oriented languages also have a notion of recursion, but it's not always central to the language's design. Similarly, object-oriented code can encapsulate the mutable state and try to get immutability as much as possible.

In today's world, even languages that we consider traditionally object-oriented, such as Java, are, in fact, becoming more and more multi-paradigm.

As a side note, this comparison might make it seem like there are only three possible paradigms: functional, object-oriented, or multi-paradigm. While these are certainly the most common, there are other paradigms, such as literate programming, logic programming, and reactive programming. As OOP is the main player in this space, and thus what most readers are familiar with, that will be a focus of comparison throughout this book.

Summary

As we have seen in this first chapter, FP is not exactly the "new kid on the block." It is a paradigm that stems from the work of Alonzo Church in the 1930s. It has seen continuous and steady investment since the 1950s with various languages pushing the paradigm further and further.

As we have also seen, FP and OOP are being combined more and more in modern languages, with Java and C# integrating ideas from the functional paradigm into their object-oriented paradigm. Go, the star of this book, takes this a step further and is a multi-paradigm language. Go gives us complete freedom to write code in whichever domain suits us best.

The core idea to remember from this chapter is that the FP paradigm will help us write code that is easier to test, read, and maintain. It reduces cognitive overhead by limiting side effects, not mutating the state of our system, and favoring small composable functions.

Finally, it is also important to remember that, although we advocate for the FP paradigm in this book, Go is multi-paradigm, and we have to choose the right paradigm for the problem we are solving.

2

Treating Functions as First-Class Citizens

As we established in the previous chapter, the core part of our functional programs will be functions. In this chapter, we are going to cover exactly why functions are powerful in languages that treat them as **first-class citizens**. Go has functions as first-class citizens out of the box, meaning we get this functionality by default. More and more languages are choosing this approach. In this chapter, we are going to see how this will allow us to create interesting constructs, which will improve the readability and test ability of our code.

Concretely, we are going to cover the following topics:

- Benefits of first-class functions
- Defining types for functions
- Using functions like objects
- Anonymous functions versus named functions
- Storing functions in data types or structs
- Creating a function dispatcher using all the previous

Technical requirements

All the examples for this chapter can be found at `https://github.com/PacktPublishing/Functional-Programming-in-Go./tree/main/Chapter2`. For this example, any Go version will work

Benefits of first-class functions

Before we talk about "first-class functions," let's first define what it means for anything to be called "first-class" in programming language design. When we talk about a "first-class citizen," we mean an entity (object, primitive, or function) for which all the common language operations are available.

These are operations such as assignment, passing it to a function, returning from a function, or storing it in another data type such as a map.

Looking at this list, we can see how all of those operations typically apply to the structs that we are defining in our language. Objects and primitives can be passed around between functions. They are often returned as the results of a function and we definitely assign them to variables. When we say that functions are first-class citizens, you can simply think of this as treating functions like objects. Their equivalence will help us create all future constructs in this book. They will lead to improved *testability*, such as by allowing us to mock functions of a struct, and improved *readability*, such as by removing large switch cases for a single function dispatcher.

Defining types for functions

Go is a statically typed language. Although, we don't have to spell out the type for each assignment – the types are there under the hood. It is effectively the compiler taking care of this for us. When we work with functions in Go, they are also implicitly assigned a type. While defining a type for a function in the way a compiler does it is a difficult task, we can use the concept of function aliases to add type safety to our code base.

When working with functions throughout the rest of this book, we will often use **type aliases**. This will help the compiler provide more readable error messages, and also makes our code more readable in general. Type aliases are, however, not just useful in the context of functions. They are a great feature of Go that is not too often used. It's also a feature that you won't easily find in other mainstream languages. So let's take a dive into what type aliases are.

In essence, a type alias does just what it says on the tin; it creates an alias for a type. It's similar to how in Unix systems you would create an alias for a command. It helps us create a new type that has all the same properties as the original type. One reason why we might want to do this is for readability, as we will see when creating aliases for functions. Another reason is to communicate our intent more clearly when we write code. For example, we could use our type system to define `CountryID` and `CityID` as an alias for `String`. Although both types are a string under the hood, they cannot be used interchangeably in code. Thus they communicate to the reader what the actual expected value is.

Type aliases for primitives

A common pattern to see in object-oriented languages is that the OO languages turn into **stringly-oriented** programming. Part of this is due to the overhead of creating a full class for a type that is, in essence, just a string. Take this example, where we have a `Person` struct, and we want to set a phone number on this struct:

```
type Person struct {
    name        string
```

```
    phonenumber string
}
func (p *Person) setPhoneNumber(s string) {
    p.phonenumber = s
}
```

In this example, which is heavily influenced by Java, we are creating a "setter-like" function that takes phonenumber as a string input and updates our object accordingly. If you are using an IDE that provides type hints for functions you are calling, it will just tell you that the setPhoneNumber function expects a string, which means any string is valid. Now, if we had a type alias, we could make that hint more useful.

So, let's make some changes and use a type alias for phoneNumber:

```
type phoneNumber string
type Person struct {
    name        string
    phonenumber phoneNumber
}
func (p *Person) setPhoneNumber(s phoneNumber) {
    p.phonenumber = s
}
```

By making this change, our type is now communicating more clearly with what our intent is, and with none of the overhead of creating a new struct just to model a phone number. We can do this, as a phone number can be thought of as essentially a string.

Using this, because a type alias is equivalent to the underlying type, is as easy as using a real string:

```
func main() {
    p := Person{
            name:        "John",
            phonenumber: "123",
    }
    fmt.Printf("%v\n", p)
}
```

Okay, great. So we have a name, which is just a string, and `phonenumber`, which is a `phoneNumber` type, which is equal to a string. So where does the benefit come from? Well, part of it is gained in communicating intent. Code is read by many more people than the original author, so we want to be as clear as possible in our code. The other part is in the error messages. Using type aliases, error messages will tell us explicitly what was expected rather than just saying a string was expected. Let's create a function that can update both `name` and `phonenumber`, and let's first use `string` for both:

```go
func (p *Person) update(name, phonenumber string) {
    p.name = name
    p.phonenumber = phonenumber
}
```

What happens when we try to compile our code? Well, we will get the following error:

```
./prog.go:26:18: cannot use phonenumber (variable of type
  string) as type phoneNumber in assignment
```

In this simple example, it doesn't do much. But as your code base expands, this ensures that all developers are thinking about the type that should be passed into a function. This lowers the risk of errors by passing invalid data to a function. An additional benefit, depending on the IDE, is that your IDE will also show you the signature. If you had a large function that takes five different types of string, your IDE might just show you *function expects input (string, string, string, string, string)*, with no clear order in which arguments need to be passed. If each string is a distinct type, this might become *name, phonenumber, email, street, country*. Especially in a language such as Go, where single-letter variable names are often used, this can create readability benefits.

To make our code work, we just have to make a small change to the function signature:

```go
func (p *Person) update(name string, phonenumber phoneNumber) {
    p.name = name
    p.phonenumber = phonenumber
}
```

This is an easy fix and amounts to just a small change, but doing it consistently makes your code convey more meaning just with the type system. In the end, types are there to convey meaning to other readers as well as the compiler.

Let's take a look at another benefit of type aliases. Let's add an `age` field to our struct with its own type alias:

```go
type age uint
type Person struct {
    name        string
```

```
    age           age
    phonenumber   phoneNumber
}
```

One thing that we cannot do with primitive types in Go such as `uint` is to attach functions to them. When we assign a type alias, however, that restriction is gone. Hence, now we can attach functions to the `age` type, which really is just attaching a function to `uint`:

```
func (a age) valid() bool {
    return a < 120
}
func isValidPerson(p Person) bool {
    return p.age.valid() && p.name != ""
}
```

In the preceding code, we are creating a `valid` function, which is bound to the `age` type. In other functions, we can now call the `valid()` function on this type with the familiar dot notation. This example is a bit trivial, but it's something that would not work on a primitive type.

If we tried to attach a function to a primitive, we would not be able to compile our program:

```
func (u uint) valid() bool {
    return u < 120
}
```

This throws the following error:

```
./prog.go:30:7: cannot define new methods on non-local type
  uint
Go build failed.
```

This alone makes type aliases quite powerful. It also means you can now extend types that are not created by you in your code base. You might be working with an external library that exposes a struct, but you want to add your own functionality to it. One way of doing that is by creating a type alias and extending it with your own functionality. While diving into this example is too in-depth for what we're exploring in this chapter, suffice it to say that type aliases are a powerful construct.

Type aliases for functions

As a function is a *first-class citizen* in Go, we can work with them like we would with any other data type. Thus, just as we can create a type alias for a variable or a struct, we can also create a type alias for a function.

Why might we want to do this? The main benefit for the reader of our code will be the clarity and readability it creates. Look at the following piece of code for a `filter` function:

```
func filter(is []int, predicate func(int) bool) []int {
    out := []int{}
    for _, i := range is {
        if predicate(i) {
            out = append(out, i)
        }
    }
    return out
}
```

This function is a good example of using functions as first-class citizens. Here, the `predicate` function is a function that is passed to the `filter` function. It is passed around in the same way in which we would typically pass around objects.

If we want to clean up this function signature, we can introduce a type alias and rewrite the filter function:

```
type predicate func(int) bool

func filter(is []int, p predicate) []int {
    out := []int{}
    for _, i := range is {
        if p(i) {
            out = append(out, i)
        }
    }
    return out
}
```

Here, you can see that the second argument now takes the `predicate` type. The compiler will translate this type to `func(int) bool`, but we can just write `predicate` throughout our code base.

Another benefit of introducing a type alias is that our error messages become more readable. Let's imagine we pass a function to `filter` that does not adhere to the `predicate` type declaration:

```
filter(ints, func(i int, s string) bool { return i > 2 })
```

Without a type alias, the error message reads as follows:

```
./prog.go:9:15: cannot use func(i int, s string) bool {…}
(value of type func(i int, s string) bool) as type func(int)
bool in argument to filter
```

That's an error message that, while being quite explicit, is quite verbose to read. With the type alias, the message will tell us what type of function we expected:

```
./prog.go:9:15: cannot use func(i int, s string) bool {…}
(value of type func(i int, s string) bool) as type predicate in
argument to filter
```

Using functions as objects

In the preceding section, we saw how to create type aliases to make our code more readable when dealing with functions. In this section, let's take a brief look at how functions can be used in the same way as objects. This is the essence of what it means to be *first-class*.

Passing functions to functions

We can pass functions to functions as in the preceding filter function:

```
type predicate func(int) bool

func largerThanTwo(i int) bool {
    return i > 2
}
func filter(is []int, p predicate) []int {
    out := []int{}
    for _, i := range is {
        if p(i) {
            out = append(out, i)
        }
    }
    return out
}
func main() {
    ints := []int{1, 2, 3}
```

```
    filter(ints, largerThanTwo)
}
```

In this example, we have created the `largerThanTwo` function, which adheres to the `predicate` type alias. Note that we don't have to specify anywhere that this function adheres to our `predicate` type; the compiler will figure this out during compile time, just like it does for regular variables. Next, we have created a `filter` function, which expects both a slice of `ints` as well as a `predicate` function. In our `main` function, we create a slice of `ints` and call the `filter` function with the `largerThanTwo` function as the second parameter.

In-line function definitions

We don't have to create functions such as `largerThanTwo` in the package scope. We can create functions inline, in the same way in which we can create structs inline:

```
func main() {
    // functions in variables
    inlinePersonStruct := struct {
            name string
    }{
            name: "John",
    }
    ints := []int{1, 2, 3}
    inlineFunction := func(i int) bool { return i > 2 }
    filter(ints, inlineFunction)
}
```

The `inlinePersonStruct` is shown in this code as an example of how the inline function compares to the inline struct definition. The code won't actually compile with this struct present as it's not used in the rest of our `main` function.

Anonymous functions

We can also create functions on the fly where they are needed. These are called *anonymous* functions as they don't have a name assigned to them. Continuing with our `filter` function, the anonymous function version of a `largerThanTwo` predicate would look like this:

```
func main() {
    filter([]int{1, 2, 3}, func(i int) bool { return i > 2 })
}
```

In the preceding example, we are both creating a slice of integers as well as the predicate function inline. Neither of them is named. The slice cannot be referenced anywhere else in that main function and neither can the function. While function definitions like these tend to make our code more verbose and can hinder readability, we will see applications of them in *Chapter 3* and *Chapter 4*.

Returning functions from functions

A core concept of any programming language is returning a value from a function. As a function is treated just like a regular object, we can return a function from a function.

In the earlier examples, our predicate largerThanTwo function always checked whether or not an integer was larger than two. Now, let's create a function that can generate such predicate functions:

```
func createLargerThanPredicate(threshold int) predicate {
    return func(i int) bool {
        return i > threshold
    }
}
```

In this example, we have created a createLargerThanPredicate function, which returns a predicate. Remember that the type predicate is just a type alias for a function that takes an integer as input and returns a bool as output. Next, we define the function that we are returning in the function body.

The function we are returning follows the type signature of predicate, and it returns true if i is larger than threshold. Notice that the i function is not passed to the createLargerThanPredicate function itself. We have defined that inline. When we call the createLargerThanPredicate function, we don't get the outcome of the predicate function, but rather we get a new function that follows the inner signature:

```
func main() {
    ints := []int{1, 2, 3}
    largerThanTwo := createLargerThanPredicate(2)
    filter(ints, largerThanTwo)
}
```

Here, in the main function, we first call the createLargerThanPredicate(2) function. This returns a new func(i int) bool function. The 2 here refers to the threshold parameter, not the i parameter.

On the next line, we can once again call the filter function with the newly created largerThanTwo function.

Returning functions from functions will be a core concept when we dive into more advanced topics such as *continuation-passing style* programming and function currying. For now, the main takeaway is that this allows us to create customizable functions on the fly. For example, we could create a series of "larger than" predicates each with its own threshold:

```
func main() {
    largerThanTwo := createLargerThanPredicate(2)
    largerThanFive := createLargerThanPredicate(5)
    largerThanHundred := createLargerThanPredicate(100)
}
```

Notice that this example won't compile, as we're not using the functions anywhere in the remainder of the `main` block. But this shows us how we can essentially "spawn" functions with one parameter fixed. Instead of creating these functions inside function blocks, we can move them up to the package-specific `var` block.

Functions in var

Continuing the preceding example, we can create a series of functions that can be used throughout our package:

```
var (
    largerThanTwo     = createLargerThanPredicate(2)
    largerThanFive    = createLargerThanPredicate(5)
    largerThanHundred = createLargerThanPredicate(100)
)
```

These "function factories" allow us to create some customized functions throughout our code. One thing to note here is that this will work inside `var` blocks, but will not compile if we move these to a `const` block:

```
const (
    largerThanTwo     = createLargerThanPredicate(2)
    largerThanFive    = createLargerThanPredicate(5)
    largerThanHundred = createLargerThanPredicate(100)
)
```

This will generate the following errors:

```
./prog.go:8:23: createLargerThanPredicate(2) (value of type
predicate) is not constant
```

```
./prog.go:9:23: createLargerThanPredicate(5) (value of type
predicate) is not constant
./prog.go:10:23: createLargerThanHundred(100) (value of type
predicate) is not constant
```

Our functions are not considered "constants" from a package perspective.

Functions inside data structures

So far, we have been creating a bunch of functions that were either defined at the top-level `var` block or inline inside a function. What if we want to store our function somewhere in the runtime memory of our application?

Well, just like we can store primitives and structs inside our runtime memory, we can store functions there as well.

Let's start off by storing our `largerThan` predicates in an array. We'll move the predicate declarations back to the `var` block and pass them to a `filter` function in our `main` function:

```
var (
    largerThanTwo     = createLargerThanPredicate(2)
    largerThanFive    = createLargerThanPredicate(5)
    largerThanHundred = createLargerThanPredicate(100)
)

func main() {
    ints := []int{1, 2, 3, 6, 101}
    predicates := []predicate{largerThanTwo, largerThanFive,
        largerThanHundred}

    for _, predicate := range predicates {
        fmt.Printf("%v\n", filter(ints, predicate))
    }
}
```

In the preceding example, we have created a "slice of predicates." The type would be `[]predicate`, and as part of the declaration, we have also pushed the three predicates we created earlier to this slice. After this line, the slice contains a reference to the three functions: `largerThanTwo`, `largerThanFive`, and `largerThanHundred`.

Once we have created this slice, we can iterate over it just like any regular slice. When we write `for _, predicate := range predicates`, the value of `predicate` takes on the value of each function we stored in the slice, sequentially. Thus, when we print the output of our filter function for each subsequent iteration, we get the following:

```
[3 6 101]
[6 101]
[101]
```

In the first iteration, `predicate` refers to the `largerThanTwo` function; in the second iteration, it becomes `largerThanFive`, and finally becomes `largerThanHundred`.

Similarly, we can store functions inside a map:

```
func main() {
    ints := []int{1, 2, 3, 6, 101}
    dispatcher := map[string]predicate{
            "2": largerThanTwo,
            "5": largerThanFive,
    }
    fmt.Printf("%v\n", filter(ints, dispatcher["2"]))
}
```

In this example, we create a map that stores predicates and associates the predicate function with a string as the key. We can then call the `filter` function and ask the map to return the function associated with the `"2"` key. This returns the following:

```
[3 6 101]
```

This pattern is quite powerful, which we'll explore later in this chapter, in *Example 1*.

Before we dive into that example, let's take a look at storing functions inside structs.

Functions inside structs

By now, it should come as no surprise that wherever we can use a data type, a function can play that role. Let's see how this plays out for structs. Let's create a struct called `ConstraintChecker`, which checks whether or not a value is in between two values.

Let's start off by defining our struct. The ConstraintChecker struct has two fields. Each field is a function of type predicate. The first function is largerThan and the second is smallerThan. These are the boundaries between which the input number should lie:

```
type ConstraintChecker struct {
    largerThan  predicate
    smallerThan predicate
}
```

Next, we create a method for this struct. The check method takes an integer input and passes this on to the largerThan and smallerThan functions, respectively. As both predicate functions return a bool, we simply check that the input returns true for both these functions:

```
func (c ConstraintChecker) check(input int) bool {
    return c.largerThan(input) && c.smallerThan(input)
}
```

Now that we have our struct and our method created, let's take a look at how we would use this struct:

```
func main() {
    checker := ConstraintChecker{
            largerThan:  createLargerThanPredicate(2),
            smallerThan: func(i int) bool { return i < 10 },
    }
    fmt.Printf("%v\n", checker.check(5))
}
```

In our main function, we first instantiate the functions. Note that we can create the ConstraintChecker struct both by providing an existing function, as we have done for largerThan, as well as by using an anonymous function as is the case for the smallerThan field.

This shows how a struct can store functions, and how these functions can be treated just like any other field in the struct. In essence, we could treat each function that is **bound** to a struct as a function that is a **field** of the struct. There are advantages to passing functions as a field versus binding it, which we will explore in more detail later, in *Example 2* in this chapter.

The main difference is that a function that is bound is essentially constant – the implementation does not change. Whereas a function passed to a field is entirely flexible. The actual implementation is unknown to our struct. We'll explore in more detail how this allows us to mock functions for testing in *Example 2*.

Example 1 – map dispatcher

One pattern that is enabled by these types of first-class functions is the "map dispatcher pattern." This is a pattern where we use a map of "key to function."

Creating a simple calculator

For this first example, let's build a really simple calculator. This is just to demonstrate the idea of dispatching functions based on a certain input value. In this case, we are going to build a calculator that takes two integers as input, an operation, and returns the result of this operation to the user. For this first example, we are only supporting the addition, subtraction, multiplication, and division operations.

First, let's define the basic functions that are supported:

```
func add(a, b int) int {
    return a + b
}
func sub(a, b int) int {
    return a - b
}
func mult(a, b int) int {
    return a + b
}
func div(a, b int) int {
    if b == 0 {
        panic("divide by zero")
    }
    return a / b
}
```

So far, this is all pretty standard stuff. We have a few functions that our calculator supports. In most cases, the result is returned instantly, but for the division function, we'll do a quick check to make sure we're not dividing by zero and panic otherwise. In a real application, we'd avoid the panic operation as much as possible, but for this example, it doesn't really have any impact. No users were harmed by having a panic in this example!

Next, let's take a look at how we'd implement the calculate function, which takes two numbers and the desired operation. We'll implement this first without considering functions as first-class citizens and use a switch statement instead to decide which operation to dispatch:

```
func calculate(a, b int, operation string) int {
    switch operation {
```

Example 1 – map dispatcher   31

```
    case "+":
            return add(a, b)
    case "-":
            return sub(a, b)
    case "*":
            return mult(a, b)
    case "/":
            return div(a, b)
    default:
            panic("operation not supported")
    }
}
```

Each branch of the switch statement performs the desired operation on our numbers and returns the result. If the options are exhausted and nothing matched the input, we panic. Each time we added a new function to our calculator, we would have to extend this function with another branch. Over time, this might not be the most readable option. So let's look at an alternative using what we've learned so far in this chapter.

First, let's introduce a type for these kinds of functions:

```
type calculateFunc func(int, int) int
```

Next, let's create a map where we can bind the string input of a user to a calculator function:

```
var (
    operations = map[string]calculateFunc{
            "+": add,
            "-": sub,
            "*": mult,
            "/": div,
    }
)
```

This map is called operations. The key of the map is the input the user will provide, which is the operations we support in our calculator. We have bound each input to a specific function call.

Now, if we want to implement the actual `calculate` function, we just have to look up the key in our map and call the corresponding function. If the requested operation does not match a key in our map, we'll panic. This is similar to the default branch in the switch-based approach:

```
func calculateWithMap(a, b int, opString string) int {
    if operation, ok := operations[opString]; ok {
        return operation(a, b)
    }
    panic("operation not supported")
}
```

This way, we can replace `Switch` statements with a map dispatcher. Also remember also that a map lookup is typically done in constant time, so this implementation of a function dispatcher is fairly efficient. It does require us to use a bit more memory to bind keys to functions, but this is negligible. With this approach, adding a new operation is a matter of adding a new entry to our map rather than extending the `switch` statement.

With the use of anonymous functions, we could also define the dispatched function inline. For example, this is how we would extend the map with bitshift functions:

```
var (
    operations = map[string]calculateFunc{
        "+": add,
        "-": sub,
        "*": mult,
        "/": div,
        "<<": func(a, b int) int { return a << b },
        ">>": func(a, b int) int { return a >> b },
    }
)
```

In this way, we can create a map dispatcher for anonymous functions. This could become rather verbose to read though, so use your best judgment when applying this.

Example 2 – mocking functions for testing

In the following example, we will take a look at mocking functions using what we have learned so far in this chapter. The application we will be building and testing is a simple to-do application. The to-do application simply allows a user to add text to a to-do, to overwrite all content.

Example 2 – mocking functions for testing   33

We won't be using an actual database, so we'll imagine that this one exists and use the filesystem and program arguments instead. Our goal will be to create tests for this application where we can mock the database interactions. To achieve this, we will use functions as first-class citizens and type aliases for code readability.

The complete example can be found on GitHub: https://github.com/PacktPublishing/ Functional-Programming-in-Go./tree/main/Chapter2/Examples/ TestingExample

Let's start by setting up our main structs. The two structs we will need are Todo and Db. The Todo struct represents the to-do item, which will contain a piece of text. The struct also contains a reference to a Db struct:

```go
type Todo struct {
    Text string
    Db    *Db
}

func NewTodo() Todo {
    return Todo{
        Text: "",
        Db:    NewDB(),
    }
}
```

In this example, we also created a "constructor" function, to ensure that users get a correctly initialized object.

We will add two functions bound to this struct: Write and Append. The Write function will override the content of the Text field, while the Append function will add content to the existing field's content. Let's also assume that any call to these functions can only be done by authorized users. As such, we'll first make a database call to figure out whether the user is authorized to perform this action:

```go
func (t *Todo) Write(s string){
    if t.Db.IsAuthorized() {
        t.Text = s
    } else {
        panic("user not authorized to write")
    }
}
func (t *Todo) Append(s string) {
    if t.Db.IsAuthorized() {
```

```
        t.Text += s
    } else {
        panic("user not authorized to append")
    }
}
```

With this in place, let's take a look at the fake database. As we want to be able to mock our database's functions in the tests that we will write later, we will leverage the concept of first-class functions. First, we'll create a Db struct. As we are only pretending that we are connecting to a real database, we won't bother with setting up the connection and having an actual database running somewhere:

```
type authorizationFunc func() bool
type Db struct {
    AuthorizationFn authorizationFunc
}
```

This is the struct definition of Db. Remember that functions can be stored as fields in a struct. And that's what's happening here, our Db struct contains a single field called AuthorizationFn. This is a reference to a function of type authorizationFunc. Remember that this is just a type alias. The compiler will actually expect a function with the func() bool signature. Thus, we are expecting a function that takes no arguments as input and returns a bool.

Now, let's create such an authorization function. As this example is self-contained, we're not interested in the overhead of having an actual database in use. For this example, assume that a user is authorized if the program arguments contain the admin string as the first argument to our program:

```
func argsAuthorization() bool {
    user := os.Args[1]
    // super secure authorization layer
    // in a real application, this would be a database call
    if user == "admin" {
        return true
    }
    return false
}
```

Notice that this function matches the function signature for the type authorizationFunc. As such, this can be stored inside the authorizationFn field of our Db struct. Next, let's create a constructor type function for our Db so we can give users a correctly initialized struct:

```
func NewDB() *Db {
    return &Db{
```

Example 2 – mocking functions for testing   35

```
            AuthorizationFn: argsAuthorization,
    }
}
```

Notice how we are passing the `argsAuthorization` function to the `AuthorizationFn` field. Whenever we are creating a database, we can thus change the implementation of `AuthorizationFn` to match our use case. We'll leverage this for unit testing later, but you could also leverage this to provide different authorization implementations, thus improving the reusability of our struct.

A handy construct to introduce here is to also create a function bound to the Db object, which will call the inner authorization function:

```
func (d *Db) IsAuthorized() bool {
    return d.AuthorizationFn()
}
```

This is a simple quality-of-life improvement. In this way, we could add code to `IsAuthorized`, which runs regardless of which implementation is chosen for the authorization function. We could add logs here for debugging, collecting metrics, handling potential exceptions, and so forth. In our case, we'll keep it as a simple function call to `AuthorizationFn`.

With this in place, let's now think about testing our code. Without mocking the `IsAuthorized` function, our tests would fail the `Write` and `Append` tests, as only authorized users can call those functions. Our test runs should not depend on the "outside world" to succeed. Unit tests should run in isolation without caring about real underlying systems (in this case, program arguments, but in a real scenario, the actual database).

So, how do we get around this? We will mock the `authorizationFn` implementation by creating a Db struct with our own `AuthorizationFn` in its place:

```
func TestTodoWrite(t *testing.T) {
    todo := pkg.Todo{
        Db: &pkg.Db{
            AuthorizationF: func() bool { return true },
        },
    }
    todo.Write("hello")
    if todo.Text != "hello" {
        t.Errorf("Expected 'hello' but got %v\n", todo.Text)
    }
    todo.Append(" world")
    if todo.Text != "hello world" {
```

```
                t.Errorf("Expected 'hello world' but got %v\n",
            todo.Text)
    }
}
```

Notice how in the setup of this test, we are manually constructing a Todo struct rather than calling the constructor-type newTodo() function. We're also manually constructing Db. This is to avoid the default implementation from running in our unit tests. Instead of using the existing function found in the code, we're providing a custom authorization function. Our custom function simply returns true for every call to IsAuthorized. This is the desired behavior in our test case, as we want to test the functionality of the Todo struct rather than that of Db. Using this pattern, we can mock core parts of our implementation. We also get the additional benefit that our structs themselves have become more flexible, as implementations can now be swapped out even at runtime.

Summary

In this chapter, we have taken a look at what first-class functions are and what type of use cases they open up to us as Go developers. We have taken a look at the equivalence between functions and objects, such as how they can be instantiated, passed around as parameters, stored inside other data structures, and returned from other functions.

We have also learned how type aliases can be used to create more readable code and to provide clearer error messages. We've seen how these can be applied to both functions as well as regular data types for structs and primitives.

In the examples, we have seen how we can create a readable function dispatcher, as well as how we can leverage first-class functions to create mocks of functions. In the next chapter, we will use what we have learned in this chapter to build higher-order functions.

3

Higher-Order Functions

In this chapter, we are going to explore the concept of function composition through higher-order functions. There are a variety of new concepts that we are introducing here, such as closures, partial application, and function currying. We will take a look at some practical examples and real-world use cases for these.

First, we will cover the core concepts of composing functions from an abstract viewpoint, and then we will combine the concepts in a practical example. Everything that we will learn here leans heavily on the concepts introduced in *Chapter 2*, where we learned what it means to treat functions as first-class citizens.

In this chapter, we will cover the following:

- An introduction to higher-order functions
- Closures and variable scoping
- Partial application
- Function currying, or how to reduce n-ary functions to unary functions
- Examples:

Technical requirements

All the examples for this chapter can be found at `https://github.com/PacktPublishing/` `Functional-Programming-in-Go./tree/main/Chapter3`. For this example, any Go version will work.

An introduction to higher-order functions

In essence, a higher-order function is any function that either takes a function as the input or returns a function as the output. Recall from the previous chapter that both of these things are made possible through the support for functions as "first-class citizens." Although it's perhaps uncommon to call them "higher-order functions," many programming languages do support these functions out of the box. For example, in Java and Python, the `map`, `filter`, and `reduce` functions are all examples of higher-order functions.

Let's create a simple example in Go. We'll have a function, A, that returns `hello,` and a function, B, that takes A as an input parameter. This is a higher-order function, as the A function is used as input to B:

```
func A() string {
    return "hello"
}
func B(a A) string {
    return A() + " world"
}
```

It is important to point out here that we're not simply passing the result of A to B – we're actually running the A function as part of the execution of B. So far, what I've shown here is not fundamentally different from anything that we saw in *Chapter 2*. Indeed, first-class functions are often demonstrated by the implementation of higher-order functions.

When they become interesting is when you start using them for partially applied computation, or when you use them to build function currying, but before we dive into these, let's look at the concept of closure first.

Closures and variable scoping

Closures are closely related to how variable scoping works in a given programming language. To fully understand how they work and how they become useful, we will first do a quick refresher on how variable scoping works in Go. Next, we'll remind ourselves of how anonymous functions work and what they are. Finally, we will take a look at what closures are in this context. This will set us up to properly understand partial application and function currying when we get to those techniques later in the chapter.

Variable scoping in Go

Variable scoping in Go is done by what is called **lexical scoping**. This means that a variable is identified and usable within the context where it was created. In Go, "blocks" are used to delineate locations in code. For example, see the following:

```
package main
import "fmt"

// location 1
func main() {
    // location 2
    b := true
    if b {
        // location 3
        fmt.Println(b)
    }
}
```

There are three locations of scope in this code:

- The first one, `location 1`, is the package scope. Our main function sits at this level of scoping.

- The next location is inside our `main` function. This is where we are defining the b Boolean.

- The third location is inside the `if` statement. In Go, and many other languages, the block is defined by curly braces.

> **Note**
>
> As a rule, variables that are defined at a "higher location" are available at a lower location, but variables defined at the lower location are not available in the surrounding higher location. In the preceding example, our code works as expected, as b is accessible from within `location 3`, even though it was defined in `location 2`.

So far, for the seasoned Go programmer, this should all pretty much be behaving as expected. Let's take a look at a few other examples of scoping. Try to figure out the output of the code prior to reading on:

Scoping example 1:

```
func main() {
    {
```

```
            b := true
    }
    if b {
        fmt.Println("b is true")
    }
}
```

What would the output be here? The right answer is… *a compilation error*. In this example, we have defined b at a different scope than the scope of the if block. Thus, we don't have access to b at this level of scoping.

Now, think about what the output would be here:

Scoping example 2:

```
func main() {
    s := "hello"
    if true {
        s := "world"
        fmt.Println(s)
    }
    fmt.Println(s)
}
```

The right answer is world hello. This might be a bit surprising. You know that you can't redeclare a variable in Go in a given scope, but, in this example, the scope inside our if statement is different from the scope of our main function. Thus, it is valid to declare a new s variable inside the if function. Do note that when using the s variable declared outside of our if statement, it has remained unchanged. This might be slightly surprising behavior. Let's change our code slightly as we jump to the third example.

Let's try to guess what the output might be of the following example:

Scoping example 3:

```
func main() {
    s := "hello"
    if true {
        s = "world"
        fmt.Println(s)
```

```
        }
        fmt.Println(s)
}
```

To point out the difference in this snippet, we have changed the first line in the `if` statement from this:

```
S := world
```

Now, it is the following:

```
S = world
```

This seemingly small difference creates the following output: `world world`. To understand this, remember that when using the `:=` syntax, we are declaring a new variable. When we only write `=`, we are redeclaring an existing variable. In this example, we are just updating the content of the `s` variable.

Now, let's make one final change to this example:

```
Scoping example 4:
```

```
func main() {
        s := "hello"
        s := "world"
        fmt.Println(s)
}
```

As you might have guessed, this code does not compile. While Go does allow us to declare variables with the same name, it only allows us to do so when they are not in the same block scope. A notable exception here is when a function returns multiple values. For example, in the following snippet, we can redeclare the error value as a return value for both `func1` and `func2`:

```
func main() {
        str1, err := func1()
        if err != nil {
                panic(err)
        }
        str2, err := func2()
        if err != nil {
                panic(err)
        }
        fmt.Printf("%v %v\n", str1, str2)
```

```
}
func func1() (string, error) {
      return "", errors.New("error 1")
}
func func2() (string, error) {
      return "", errors.New("error 2")
}
```

In the preceding snippet, the `err` value gets redeclared even though we are using the `: =` syntax. This is commonly encountered in Go as the error values bubble up from each function to an eventual parent method that handles multiple errors.

It is important to remember how scoping works and the significance of the curly braces to delineate blocks, as well as to remember the difference between introducing a new variable versus simply redeclaring an existing one. With this out of the way, we have enough background knowledge to jump into variable scoping when using functions inside functions.

Capturing variable context in functions (closures)

In the previous chapter, we saw that each time we encountered curly braces, a new variable scope was introduced. This happens when we declare a function, branch into an `if` statement, introduce a `for` loop, or simply place curly braces anywhere in a function, as in our first scoping example. We also saw in *Chapter 2* that we can create functions inside functions – and, as you might have guessed, this creates a new scope yet again.

For the remainder of this chapter, we will frequently use anonymous functions. Remember that an anonymous function is essentially a function declaration without an identifier attached to it. This is the general template that we are using:

```
// location 1
func outerFunction() func() {
      // location 2
      fmt.Println("outer function")
      return func() {
            // location 3
            fmt.Println("inner function")
      }
}
```

In this example, I have denoted the three variable scoping locations as well. As you can see, `location 3`, which is part of the anonymous function, is scoped at a lower level than `location 2`. This is a critical reason why closures work. Defining a new function does not automatically create a top-level scope. When we define a function inside another function, this new function scopes variables at a lower level than where it was introduced.

Also, note that outerFunction is a higher-order function. Although we don't take a function as input, we are returning a function as output. This is a valid characteristic of higher-order functions.

Now, let's say specifically what we mean by a closure. A closure is *any inner function that uses a variable introduced in the outer function* to perform its work. Let's make this more concrete by looking at an example.

In this example, we are going to create a function that creates a greeting function. Our outer function will be the function that determines the greeting message to show. The inner function will ask for a name as input and return the greeting combined with the name:

```
func main() {
    greetingFunc := createGreeting()
    response := greetingFunc("Ana")
    fmt.Println(response)
}

func createGreeting() func(string) string {
    s := "Hello "
    return func(name string) string {
        return s + name
    }
}
```

In the preceding example, we are using a closure. The anonymous inner function references the outer variable, s, to create the greeting. The output of this code is Hello Ana. What is important here is that, although the s variable went out of scope once the createGreeting function ended, the variable content is actually captured inside the inner function. Thus, after we called greetingFunc in our main function, the capture was fixed as Hello. Capturing a variable inside an inner function is what is meant when we talk about closures.

We can make this function more flexible by accepting the greeting string as an input parameter to the createGreeting function so that we get the following:

```
func createGreeting(greeting string) func(string) string {..}
```

This small change brings us to the start of the next topic: partial applications.

Partial application

Now that we understand closures, we can start thinking about partial application. The name "partial application" quite explicitly tells us what is happening – it is a function that is partially applied. This is perhaps still a bit cryptic. A partially applied function is taking a function that takes *N* number of arguments and "fixing" a subset of these arguments. By fixing a subset of the arguments, they become set in stone, while the other input parameters remain flexible.

This is perhaps best shown with an example. Let's extend the createGreeting function that we built in the previous section of this chapter:

```
func createGreeting(greeting string) func(string) string {
    return func(name string) string {
        return greeting + name
    }
}
```

The change we have made here is to have the greeting passed as an input to the createGreeting function. Each time that we call createGreeting, we are effectively creating a new function, which expects name as input but has the greeting string fixed. Let's create a few of those functions now and use them to print the output:

```
func main() {
    firstGreeting := createGreeting("Well, hello there ")
    secondGreeting := createGreeting("Hola ")
    fmt.Println(firstGreeting("Remi"))
    fmt.Println(firstGreeting("Sean"))
    fmt.Println(secondGreeting("Ana"))
}
```

The output of running this function is as follows:

```
Well, hello there Remi
Well, hello there Sean
Hola Ana
```

In this example, we fix the first parameter of the firstGreeting function as Well, hello there, while for the secondGreeting function, we have fixed the value as Hola. This is partial application – when we create the function to greet users with a name, part of this function has already been applied. In this case, the greeting variable was fixed, but you can fix any subset of the arguments of a function – it's not limited to just one variable.

Example: DogSpawner

In this example, we are going to tie everything that we have learned so far together. For this example, we are going to create DogSpawner. You can imagine that this could be used in the context of creating a game or another application for which you'd be maintaining information on dogs. As in our other examples, we are going to trim this down to the bare essentials and we won't be making an actual game. What we are going to do in this example, however, is leverage what we've learned in previous chapters and tie it all together with clean functional code.

From a high-level point of view, our application should support dogs of multiple breeds. The breeds should be easily extensible. We also want to record the gender of the dog and give the dog a name. In our example, imagine that you'd want to spawn many dogs, so there would be a lot of repetition of types and genders. We'll leverage partial application to prevent the repetitiveness of those function calls and improve the code readability.

First, we will start by defining the types that we'll need for this program. Remember from the first chapter that we can use the type system to give you more information about what is happening in the code:

```
type (
    Name          string
    Breed         int
    Gender        int
    NameToDogFunc func(Name) Dog
)
```

Notice that we can use a type block, similar to how we can use a var or const block. This prevents us from having to repeat the type Name string structure. In this type block, we have simply chosen Name to be a string object, Breed and Gender to be int objects, and NameToDogFunc is a function that takes in a given Name and returns a given Dog as a result. The reason we chose int objects for Breed and Gender is that we'll construct those using Go's equivalent of an Enum definition. We'll go ahead and populate these enums with some values:

```
// define possible breeds
const (
    Bulldog Breed = iota
    Havanese
    Cavalier
    Poodle
)
// define possible genders
const (
```

```
        Male Gender = iota
        Female
)
```

As you can tell from the preceding example, the default `iota` keyword works out of the box with the types that we have defined. Once again, this shows that our type aliases compile down to the underlying type, in this case, the `int` type for which `iota` is defined. You could merge the two `const` blocks in this example into a single block, but when dealing with enumerations, the code remains more readable when each block serves a single purpose.

With these constants and types in place, we can create a struct to represent our Dog:

```
type Dog struct {
    Name    Name
    Breed   Breed
    Gender  Gender
}
```

It's a bit repetitive in this struct, as the names of our variables are identical to the type. For this example, we can keep it lightweight and don't have to add any more information to our Dog. With this in place, we have everything we need to start implementing our partially applied functions, but before we get to that, let's look at how we'd create Dog structs without partially applied functions:

```
func createDogsWithoutPartialApplication() {
    bucky := Dog{
            Name:   "Bucky",
            Breed:  Havanese,
            Gender: Male,
    }
    rocky := Dog{
            Name:   "Rocky",
            Breed:  Havanese,
            Gender: Male,
    }
    tipsy := Dog{
            Name:   "Tipsy",
            Breed:  Poodle,
            Gender: Female,
    }
}
```

In the preceding example, we have created three dogs. The first two are both male Havanese dogs, so we had to repeat the `Breed` and `Gender` information there. The only thing that's unique between those two would be the name. Now, let's create a function that allows us to create `DogSpawner` of various gender and breed combinations:

```
func DogSpawner(breed Breed, gender Gender) NameToDogFunc {
    return func(n Name) Dog {
        return Dog {
            Breed:  breed,
            Gender: gender,
            Name:   n,
        }
    }
}
```

The preceding `DogSpawner` function is a function that takes `Breed` and `Gender` as input. It returns a new function, `NameToDogFunc`, which takes `Name` as input and returns a new `Dog` struct. This `DogSpawner` function thus allows us to create new functions where the dog's breed and gender are already partially applied, but the name is still expected as input.

Using the `DogSpawner` function, we can create two new functions, `maleHavaneseSpawner` and `femalePoodleSpawner`. These functions will allow us to create male Havanese dogs and female poodles, by only providing a name for our dogs. Let's go ahead and create two new functions in the package-scoped `var` block:

```
var (
    maleHavaneseSpawner = DogSpawner(Havanese, Male)
    femalePoodleSpawner = DogSpawner(Poodle, Female)
)
```

After this definition, the `maleHavaneseSpawner` and `femalePoodleSpawner` functions are available anywhere in that package. You could also expose them as public functions that anyone using the package has access to. Let's demonstrate in our `main` function how these functions could be used:

```
func main() {
    bucky := maleHavaneseSpawner("bucky")
    rocky := maleHavaneseSpawner("rocky")
    tipsy := femalePoodleSpawner("tipsy")

    fmt.Printf("%v\n", bucky)
    fmt.Printf("%v\n", rocky)
```

```
        fmt.Printf("%v\n", tipsy)
}
```

In this `main` function, we can see how we can leverage the partially applied functions. We could have created a function to create dogs, such as `newDog(n Name, b Breed, g Gender) Dog{}`, but this would still have led to a lot of repetition in creating our dogs, as follows:

```
func main() {
        createDog("bucky", Havanese, Male)
        createDog("rocky", Havanese, Male)
        createDog("tipsy", Poodle, Female)
        createDog("keeno", Cavalier, Male)
}
```

While still decently readable with only three parameters, more parameters will significantly impair readability. We'll show this in the last example of this chapter after we've discussed function currying.

Function currying, or how to reduce n-ary functions to unary functions

Function currying is often mistaken for partial application. As you will see, function currying and partial application are related but not identical concepts. When we talk about function currying, we are talking about transforming a function that takes a single argument to a sequence of functions where each function takes exactly one argument. In pseudocode, what we are doing is transforming a function such as the following into a sequence of three functions:

```
func F(a,b,c): int {}
```

The first function, (`Fa`), takes the `a` argument as input and returns a new function, (`Fb`), as output. (`Fb`) takes b as input and returns an (`Fc`) function. (`Fc`), the final function, takes c as input and returns an `int` object as output:

```
func Fa(a): Fb(b)
func Fb(b): Fc(c)
func Fc(c): int
```

This is done by leveraging the concept of first-class citizens and higher-order functions once again. We'll be able to achieve this transformation by returning a function from a function. The core feature that we'll achieve from this is more composable functions. For our purposes, you can think of this as partial application applied to single arguments.

One thing to note here is that in other programming languages such as Haskell, function currying plays a much more important role than here in our Go examples. Haskell (which is named after Haskell Curry), transforms each function into a curried function. The compiler takes care of that, so you're not generally aware of this as a user. The Go compiler does no such thing, but we can still manually create functions in such a way. Before we dive into larger end-to-end examples, let's take a quick look at how we'd transform the previous pseudocode into functioning Go code.

Without currying, our function would look like this:

```go
func threeSum(a, b, c int) int {
    return a + b + c
}
```

Now, with currying, the same example would translate to this:

```go
func threeSumCurried(a int) func(int) func(int) int {
    return func(b int) func(int) int {
        return func(c int) int {
            return a + b + c
        }
    }
}
```

When calling them in the main function, these return the same result. Notice the difference in syntax between the two calls in the main function:

```go
func main() {
    fmt.Println(threeSum(10, 20, 30))
    fmt.Println(threeSumCurried(10)(20)(30))
}
```

It should go without saying that the curried version of this function is way more complicated to read and comprehend than the uncurried function. This ties back to what I mentioned in the first chapter – you should leverage functional concepts where they make sense. For this simple example, it didn't make sense but it does demonstrate the point of what we are trying to do. The real power of function currying only comes in handy when we also decide to combine it with partial application to create flexible functions. To show how this works, let's dive into an example.

Example: function currying

In this example, we are going to extend the functionality of the DogSpawner example that we've built to demonstrate partial application. If we look at the main DogSpawner code for that application, we can tell that we are almost using a unary function:

```
func DogSpawner(breed Breed, gender Gender) NameToDogFunc {
    // implementation
}
```

That gets us close, but no dice. To be a properly curried function, DogSpawner can only take one parameter. In essence, we are going to create a sequence of three functions that take the successive arguments to create Dog, DogSpawner(Breed)(Gender)(Name). If we implement this function in Go, we get the following code:

```
func DogSpawnerCurry(breed Breed) func(Gender) NameToDogFunc {
    return func(gender Gender) NameToDogFunc {
        return func(name Name) Dog {
            return Dog{
                    Breed:  breed,
                    Gender: gender,
                 Name:    name,
            }
        }
    }
}
```

The way to read this is that DogSpawnerCurry is a function that takes breed as input. It returns a function that takes gender as input, which, in turn, returns a function that takes name as input and returns Dog. This is a bit complex to read, but you get the hang of it. This is also where type aliases come in handy. Without a type alias, this would be even more verbose, which would hinder reading and make it more error-prone to write:

```
func DogSpawnerCurry(breed Breed) func(Gender) func(Name) Dog {
    return func(gender Gender) func(Name) Dog{
        return func(name Name) Dog {
            return Dog{
                    Breed:  breed,
                    Gender: gender,
                    Name:    name,
```

```
                    }
                }
            }
        }
```

Now that we have covered the three main themes of this chapter, let's take a look at some further examples to demonstrate these techniques.

Example: server constructor

In this first example, we are going to leverage what we've learned so far to create flexible constructors for data types. We will also see how we can create constructors with default values of our choosing.

In our setup, a `Server` struct is a simple struct that has a set number of maximum connections, a transport type, and a name. We won't be building an actual web server, but rather, we are demonstrating the concepts with only a small amount of overhead. What we want to do in this example is to focus on the core ideas, which you can then apply anywhere you see fit. Our server only has three configurable parameters, but you can imagine that this benefit is more pronounced when there are more parameters to configure.

As always, we are going to start by defining the custom types of our application. To keep it lightweight, I'm defining two of them – `TransportType`, which is an `int` type to be used as an enumeration, and a type alias for `func(options) options`. Let's also set some values for `TransportType`:

```
type (
    ServerOptions func(options) options
    TransportType int
)
const (
    UDP TransportType = iota
    TCP
)
```

Now that we have this, let's get our structs in place – the two structs that we will be using as `Server` and `options`:

```
type Server struct {
    options
```

```
}
type options struct {
    MaxConnection int
    TransportType TransportType
    Name          string
}
```

In the example here, we have embedded `options` without declaring a new name for the field. This is achieved in Go by simply writing the type of struct that you want to embed. When doing so, the `Server` struct will contain all the fields that the `options` struct has. It's a way to model object composition in Go.

This might look a bit peculiar and warrants some further investigation. In a more typical setup, you might have the `Server` struct contain the variables that we have placed inside the `options` struct. The main reason for using the `options` struct and embedding it inside `Server` is to use this as a configuration for our server that we want users to provide. We don't want users to provide data that is not contained in this struct, such as the `isAlive` flag. This clearly separates concerns, and it will allow us to build the next higher-order functions and partial application layers on top of it.

The next step is creating a way for us to configure the `options` struct through multiple function calls. For each variable inside the `options` struct, we are creating a higher-order function. These are functions that take in the parameter to be configured, and return a new function, `ServerOptions`:

```
func MaxConnection(n int) ServerOptions {
    return func(o options) options {
    o.MaxConnection = n
        return o
    }
}
func ServerName(n string) ServerOptions {
    return func(o options) options {
        o.Name = n
        return o
    }
}
func Transport(t TransportType) ServerOptions {
        return func(o options) options {
                o.TransportType = t
                return o
```

```
        }
    }
```

As you can see in the preceding three functions (`MaxConnection`, `ServerName`, and `TransportType`), we are using a closure to build this configuration. Each function takes in a struct of the `options` type, changes the corresponding variable, and returns the same `options` struct with the change applied. *Notice that these functions only change their corresponding variable, and everything else in the struct remains untouched.*

Now that we have this, we have everything in place to start constructing our server. For our constructor, we'll write a function that takes a variadic argument list of `ServerOptions` as our input. Remember that these inputs are really other functions. Our constructor is a higher-order function that takes functions as input and returns a server as output. Thus, when we iterate over our `ServerOptions`, we get a series of functions that we can call. We'll create a default struct of `options` to pass to these functions:

```
func NewServer(os ...ServerOptions) Server {
    opts := options{}
    for _, option := range os {
        opts = option(opts)
    }
    return Server{
        options: opts,
        isAlive: true,
    }
}
```

In the code here, you can see how our `Server` is finally built based on the `options` struct. We're also setting the `isAlive` flag to `true`, as this is not something the user could input.

Great, we have everything in place to start creating servers – so how do we go about that? Well, our constructor is a bit different from other constructors that you might have seen. Rather than taking variables such as primitives or structs as input, we are going to pass functions as input. Let's demonstrate in the `main` function how we can call this constructor:

```
func main() {
    server := NewServer(MaxConnection(10),
ServerName("MyFirstServer"))
    fmt.Printf("%+v\n", server)
}
```

As you can tell, we call the `MaxConnection(10)` function inside the constructor. The output of this function is not simply a struct; the output is `function(options) options`. When running this code, we get the following output:

```
{options:{MaxConnection:10 TransportType:0 Name:MyFirstServer}
    isAlive:true}
```

Great – now, we have a quite flexible constructor. If you notice in the output, we get `TransportType: 0` as output, even though we did not configure this in our `options` struct. This is because Go uses a sane default zero value for its primitive types. One thing our current constructor setup allows us to do is to create default values that we set ourselves with only minor changes to our code. Let's update the `NewServer` function to use TCP (`TransportType: 1`) as the default value:

```
func NewServer(os ...ServerOptions) Server {
        opts := options{
                TransportType: TCP,
        }
        for _, option := range os {
                opts = option(opts)
        }
        return Server{
                options: opts,
                isAlive: true,
        }
}
```

In the example, the only change we made was to add `TransportType: TCP` to the initialization for our `options`. Now, if we run the same main code again, we get the following output:

```
{options:{MaxConnection:10 TransportType:1 Name:MyFirstServer}
    isAlive:true}
```

This is how easy it is to create our own default values when a user does not provide any. As this example shows, we can easily use functional programming concepts to build flexible functions such as constructors and achieve functionality that is not natively present in Go. In some languages, such as Python, you can set default values for a function when the user does not provide them. Now, we can do the same thing using the `options` struct for our server.

Summary

In this chapter, we covered three things: closures, partial application, and currying. By using closures, we learned how we can share the context of variables between outer and inner functions. This allowed us to build flexible applications, such as the final "constructor" example. Next, we learned how to use a partially applied function to fix certain arguments to an n-ary function. This shows us how we can create default configurations for functions, such as how we created a `HavaneseSpawner` option in our example. Finally, we learned about function currying and how this relates to partial application. We showed how we can extend our partial application example by transforming each function into unary function calls. All three techniques have allowed us to create more composable and reusable functions.

Up until now, we have not been concerned with function purity and have played a bit fast and loose with the state of our system. In the next chapter, we are going to talk about what it means for functions to be pure, how we can encapsulate side effects, and what benefits this brings for writing testable code.

4

Writing Testable Code with Pure Functions

When you read about functional programming, quite often, what is meant is "pure" functional programming. As we touched on in the first chapter, this is not a strict requirement of functional programming or functional languages. If you decide to pick up a functional programming language, the chances are pretty high that you'll pick up a language such as Haskell or Elm. If so, you would have chosen two purely functional languages and might have coupled your understanding of *pure functional* with *functional*. On the other hand, if you had picked up a language such as Lisp, Clojure, or Erlang, you would have picked a functional language that is impure yet still functional.

In this chapter, we will address the following topics:

- What exactly is purity?

- Why should purity matter?

- How do we create pure functions?

- Learning how unit testing is impacted by writing pure functions

Technical requirements

For this chapter, any version of Go after Go 1.12 can be used. You can find the complete examples at https://github.com/PacktPublishing/Functional-Programming-in-Go./tree/main/Chapter4.

What is purity?

When talking about a purely functional programming language, we are talking about a language in which each function adheres to these properties:

- Does not generate any side effects
- Returns the same output when providing the same input (idempotence)

This means that our functions are completely deterministic.

The best way forward might be to demonstrate what we are talking about by showing some examples. So, in this section, we'll take a look at two functions, a pure one and another which is impure. Then, we'll talk a bit more about the properties of such functions and their importance to the programs that we are writing.

Demonstrating pure versus impure function calls

A simple example of this would be an addition function. This is a function that takes two integers as input and returns the sum as the output:

```
func add(a, b int) int {
    return a + b
}
```

When we call this function with the same inputs, we will get consistent output. Thus, no matter how many times I call the add(10,5) function, the code will always return the same output: 15. This is pretty much as simple as it gets when creating a pure function. We did not use any state outside of our function to determine the answer, nor did we update anything outside of our function.

Next, let's look at an example of an impure function, whose output is always random:

```
func rollDice() int {
    return rand.Intn(6)
}
```

When calling the rollDice function, the output is not consistent. If it were consistently outputting the same number, it would be a pretty bad randomization function. If we were to call the rollDice function five times, we'd get five different outputs:

```
func main() {
    for i := 0; i < 5; i++ {
            fmt.Printf("dice roll: %v\n", rollDice())
```

```
    }
}
```

This would result in the following output:

```
dice roll: 5
dice roll: 3
dice roll: 5
dice roll: 5
dice roll: 1
```

Referential transparency

One property that helps us think about pure functions is the property of *referential transparency*. Both in mathematics and computer science, a function is said to be referentially transparent if you can replace the function call with its output, without changing the result of the program. In mathematics, it's easy to see why that is true. If we work out any formula, we can essentially substitute part of an equation for its result, without changing the result. For example, take the following equation:

```
X = 1 + (2 * 2)
```

The result is 5. We could have gotten the same result had we replaced the multiplication with its result, like so:

```
X = (1 + 4)
```

This property is what we mean by referential transparency. All mathematical operations have this property, and many of us have leveraged this property when working out equations in our algebra, calculus, or other mathematics classes.

Let's jump back to the realm of software engineering to explore this further. In a programming language, referential transparency means that a function call can be replaced with its result. If we apply this same test to our `add` function, which we wrote earlier, we can see how this is true. Let's demonstrate this with a small piece of code:

```
func main() {
    fmt.Printf("%v\n", add(10, add(10, 5)))
    fmt.Printf("%v\n", add(10, 15))
}

func add(a, b int) int {
```

```
        return a + b
}
```

In this example, we have replaced one of the add functions with its result. And sure enough, the output of our program remained identical and functionally correct. You might think this is obvious, but there are plenty of functions that we rely on for which this is not true. Let's introduce another function that breaks this property. We'll keep it simple and create a program that tells us the current time:

```
func main() {
    fmt.Printf("%v\n", time.Now())
}
```

In this snippet, we are using the time.Now function. There's not a single value that you can replace this function call with while guaranteeing your program is functionally equivalent and correct. If we were to hardcode the current time, it would be wrong by the time your program has compiled and when it's run.

To expand on this, let's take a look at a larger example than just the time.Now function. In the following piece of code, let's imagine we are writing a function to select the starting player of a game. We'll use a simple type alias from Player to string instead of making a complete struct. As this is a game, we want our starting player to be randomly selected on each run of our program:

```
type Player string
const (
    PlayerOne Player = "Remi"
    PlayerTwo Player = "Yvonne"
)
func selectStartingPlayer() Player {
    randomized := rand.Intn(2)
    switch randomized {
    case 0:
            return PlayerOne
    case 1:
            return PlayerTwo
    }
    panic("No further player available")
}
```

In the preceding code, we are breaking the referential transparency requirement of our code since there is no way to replace this function call with a single value while maintaining an equivalent outcome of our program. The preceding code is also not testable. Think about this for a second – how would you write a unit test for this function? This would prove impossible to do in the code's current state and would require some refactoring. We will show you how we could refactor this code and make it testable later in this chapter, but you can find it on GitHub at `https://github.com/PacktPublishing/Functional-Programming-in-Go./tree/main/Chapter4/TestableCode`.

Idempotence

Another property of pure functions is that they are idempotent. This means that no matter how many times the function is executed, it will always return the same output, assuming that the input arguments have remained the same. In the preceding example, the `add` function is always returning the same sum of two numbers provided the same input. On the other hand, the `time.Now` function is not (nor would that have been the desired behavior).

You might be familiar with idempotence as it also shows up when you are implementing a REST service or dealing with HTTP calls in general. When implemented correctly, the `GET`, `HEAD`, `PUT`, and `DELETE` methods should be idempotent. A notable exception is the `POST` method.

Statelessness

A pure function should not depend on any state of the system. This means that neither the input nor the output should change the state. Web requests are often said to be stateless; each request can run independently of the other and still generate the same result. In Go terms, this also means that our function should not depend on things such as global variables, files on our filesystem, or general I/O operations.

Side effects

The properties mentioned previously tie together in creating functions that are free of side effects. A side effect is any operation that your function does that changes the state of your system. In the next chapter, we'll dive deeper into what it means for the state to be immutable at the `struct` level. In this chapter, we'll consider the state to mean the system within which your program is operating.

Why does purity improve our code?

So far, we have looked into some properties of purely functional code. We've also seen some examples of both pure and impure functions. Now, let's look at what benefits we can expect from writing pure functional code.

Increases the testability of our code

When writing pure functions, your functions will be easier to test. This is a consequence of them being both idempotent and stateless:

- **Idempotent**: Run functions any number of times and get the same result
- **Stateless**: Each function will run independently of the state of the system

For idempotence, it's easy to see how this would be true. In our test suite, if functions were to return different outputs for the same inputs, it would be hard to write tests for that function. After all, if you can't predict the output of a certain function, you can only guess what value you should be testing for. The benefit of it being stateless might not be immediately obvious. This comes down to our test suite not being able to run in the same environment as our production system. Thus, if we were relying on the state of the system in some way, we would have to guarantee that our test state replicates the production state at the moment in which the function is called. Let's demonstrate this with an example.

Recall from earlier in this chapter, when we created a function to select a random player for a game? Let's refactor this code into something more testable. There are two changes we need to make – first, we need to make the function deterministic. It sounds like this breaks the randomization, and it does, but we'll show how we can work around that soon. The second change we will make is to remove any side effects. In our first example, we had a `panic` function in case the randomization function returned an integer above 1. We'll replace that panic by returning a tuple from our function containing `(Player, error)`, following the common error handling idiom in Go. With these changes, our new function looks like this:

```
func PlayerSelectPure(i int) (Player, error) {
    switch i {
    case 0:
            return PlayerOne, nil
    case 1:
            return PlayerTwo, nil
    }
    return Player(""), fmt.Errorf("no player matching input:
        %v", i)
}
```

With these changes in place, our function is now deterministic. For each input, we always generate the same output, like so:

```
PlayerSelectPure(0) = PlayerOne, nil
PlayerSelectPure(1) = PlayerTwo, nil
PlayerSelectPure(n > 1) = Player{}, error
```

Notice how in the last case, whereby n is larger than one, we are not simply returning `nil` and an error. This bears some explanation. The gist of it is that we'll try to avoid using pointers in our code as much as possible. And, in Go, if you're not working with pointers, you cannot represent `nil`. Exactly why we avoid this and what the implications are will be explained in detail in the following chapter, *Chapter 5*.

Now that we have seen what the expected output is for each case, and that we agree on this function being pure, we can write a test case to confirm that the output matches what we expect:

```
func TestPlayerSelectionPure(t *testing.T) {
    selectPlayerOne, err := PlayerSelectPure(0)
    if selectPlayerOne != PlayerOne || err != nil {
        t.Errorf("expected %v but got %v\n", PlayerOne,
            selectPlayerOne)
    }

    selectPlayerTwo, err := PlayerSelectPure(1)
    if selectPlayerTwo != PlayerTwo || err != nil {
        t.Errorf("expected %v but got %v\n", PlayerOne,
            selectPlayerTwo)
    }

    _, err = PlayerSelectPure(2)
    if err == nil {
        t.Error("Expected error but received nil")
    }
}
```

Everything that is happening in the preceding code is pretty straightforward. For each valid input (0 and 1), we confirm that the first or second player is returned, respectively. For an input higher than 1, we confirm that an error is thrown. Technically, you could extend this unit test to exhaustively test all possible integer inputs and confirm that an error is thrown for each of them. That might be a tad too exhaustive for this simple function, though.

With this, only one thing remains to be addressed: our code is no longer selecting a random player, but rather it is expecting an integer input and returning a deterministic value. You might notice that we have merely shifted the problem, as the random selection function still needs to be present somewhere. This is correct. If we were to look at how we would be using this code in an actual game, we might find code like this:

```
func main() {
    random := rand.Intn(2)
```

```
        player.PlayerSelectPure(random)
        // start the game
}
```

Here, we can see a recurring pattern as we aim to improve the purity of our code. The strategy will be to limit the places where side effects and non-determinism can occur. When you shift the way you think about structuring your code to preferring function purity and isolating the locations where you break it, you might end up with 90% of pure code and 10% impure code. Sure, you're not 100% purely functional, but we're programming in Go and we can forgive ourselves for the 10% impure code. As we've explored in quite some detail, purely functional programming is a subset of functional programming. Also, there's no pure functional programming police that will hunt you down if you write an impure function.

Does this mean it's impossible to be entirely pure? Well, not quite. After all, there are purely functional programming languages such as Haskell that can be used in real-world production environments. The way they deal with these impure functions is by using a form of encapsulation, known as **monads**. While it is possible to create monads in Go, it might cause more friction than necessary, hence why I advocate for embracing the idea of functional rather than pure functional code. For fun and to be extensive in our exploration of purely functional code, we will take a look at monads in the next chapter.

Increases the confidence in our code

While this goes hand in hand with improved testability, the improved confidence in your code goes beyond that. When dealing with impure functions and states, your program is harder to comprehend. If you work in a sufficiently complex system that has impure functions, and state mutations such as through global variables, it becomes harder to reason about. Imagine you are working in such a complex system, and a user reports a bug. If the system is mutable, you'd need to get a full understanding of what the entire system looked like at the moment the bug appeared just to begin to debug it. This can lead to many painful and wasted hours of debugging. There's a popular notion, called the **Heisenbug**, which is a consequence of this. In this case, if the function that caused the bug depended on the state of the system, you might need to repeat the exact steps the user did just to replicate the bug.

An additional benefit is that our code becomes easier to debug. When debugging a program, any sufficiently advanced debugger will also show the state of your system during debugging. It will tell you what values were held in memory by the various parts of your program. That's a great tool and can help you find bugs and eliminate them. But what if your program just doesn't depend on such a state? This would eliminate the need for a crux such as the advanced debugger.

You could look at a single function, and reason about what it's doing without having to also keep in mind what the rest of the system would look like at the moment of execution. Humans are bad at keeping things in our "working memory;" we can juggle about 7 +- 2 things at any given moment. If we optimize and try to make our program understandable by most humans, we'd have to limit the state variables to just 5. This is ignoring the fact that our function probably has some variables as well. As such, we quickly exceed the upper bound of our human memory limits.

Improved confidence in function names and signatures

Another great benefit of improving the readability and understandability of your code is that you can suddenly gain extra confidence in your functions. Imagine you are reading a code base, and you encounter the following piece of code:

```
func main() {
    i := 1
    for i < 10 {
        i = add1(i)
        fmt.Printf("%v,", i)
    }
}
```

What would the output be? You may naturally assume that add1 is a pure function, and the output would be as follows:

```
1,2,3,4,5,6,7,8,9,10,
```

But, you would be wrong. The actual output is as follows:

```
1, 2, 3, 4, 5, 6, panic: can not increment any more
goroutine 1 [running]:
main.add1(...)
    /tmp/sandbox1318301126/prog.go:17
main.main()
    /tmp/sandbox1318301126/prog.go:10 +0xa5
Program exited.
```

To understand why, let's take a look at the add1 function:

```
func add1(input int) int {
    if input != 0 && input > rand.Intn(input) {
        panic("can not increment any more")
    }
    return input + 1
}
```

In the preceding function, we can see that the add1 function is impure. It is not deterministic, as the outcome of each run depends on a random number being generated. Plus, it also creates a side effect. Each time a function has a panic statement in it, that statement produces a side effect outside of the normal outcome of your function. This was a bit of a contrived example, but it shows that when working in an environment where functions can contain side effects and are not idempotent, you lose some trust in the function signature itself.

Safer concurrency

One of the selling points of Go, and a feature that sets it apart from many mainstream languages, is how easily it handles concurrency. With Go, it is incredibly easy to spin up multiple threads and have them work in parallel. This happens through the **channels** and **goroutines** concepts. There is much to be said about how concurrency works in Go, enough to deserve its entire book. What we will focus on here briefly is the correctness aspect of concurrency. Is it true that spinning up goroutines and processing in parallel is much easier in Go compared to, say, Java? What is not true is that it's easier to write correct concurrent code.

Let's take a look at some concurrent code. In this example, we will create a slice of integers and append to it in the `addToSlice` function. In our `main` function, we will push an integer to the slice:

```go
var (
    integers = []int{}
)

func addToSlice(i int, wg *sync.WaitGroup) {
    integers = append(integers, i)
    wg.Done()
}

func main() {
    wg := sync.WaitGroup{}
    numbersToAdd := 10
    wg.Add(numbersToAdd)

    for i := 0; i < numbersToAdd; i++ {
        go addToSlice(i, &wg)
    }

    wg.Wait()
    fmt.Println(integers)
}
```

Think about this program for a second and try to guess what the output would be. The correct answer is that this program's output is non-deterministic. We are running multiple threads in which we append to our slice, and at the end, we call `wg.Done()`. When working with these waitgroups, we pass along several threads to wait for. This is done in `wg.Add(numbersToAdd)`. Each time `wg.Done()` is called, the number of threads to wait for decrements by one. As we are working on a shared slice of integers in this example, it is impossible to predict exactly what that slice looked like when it was performing the `add` operation in the final thread. This means that our output could be all numbers randomly ordered such as `[9 0 1 2 3 4 5 6 7 8]`, but it's equally possible that the output would just be `[4 9 0 1 2]`. Having mutable data sources in concurrent functions is a recipe for disaster, and leads to some pretty hard-to-trace bugs.

So, as you can see from this small snippet, it was incredibly simple to spin up multiple threads, but not quite so simple to avoid bugs in our code. Pure functions can help with this. Remember that when a function is pure, the same input always generates the same output without causing any side effects.

In this example, our side effect was mutating the slice, which is not thread-safe in Go. The program won't crash but the result will be random. If we push pure functional programming to the extreme, we will eliminate all such impure functions, and in doing so, we could run all our functions infinitely in parallel without causing any trouble.

> **Note**
> In practice, there are ways to avoid this from happening using a mutex. Some libraries take care of parallelism and thus abstract away some of the complexity.

When not to write pure functions

So far, we have seen what pure functions are and what kind of advantages pure functions can provide. But we should at least spend a bit of time thinking about occasions where we might want to sacrifice function purity. Now, if you ask this question to "purists," the answer to this question is probably along the lines of: "Never, nunca, jamais." This is fine, and some languages make it pretty easy to write functional code without ever having to sacrifice function purity. But, let's take a look at a few examples where it makes sense to sacrifice some function purity. Now, before we dive into these examples, let me be the first to acknowledge that all of these supposed problems are circumventable. And yes, a language such as Haskell handles this mostly gracefully.

But we are not programming in Haskell; we are programming in Go. And while Go allows us to write purely functional code if we wish to do so, some things are just easier to implement by temporarily forgiving ourselves for our sin of writing impure code.

Input/output operations

Think about the implications of totally eradicating side effects from your code. If we say that we are writing purely functional code and have eliminated all side effects, we have also eliminated part of what generates value for our users. Any time we are getting input from a user or displaying input to a user in some way, it is technically a side effect. Any time we are storing data in local storage, or uploading to a server somewhere, we are producing a side effect. Many applications will take some type of input, and many will also generate some kind of output.

Non-determinism can be desired

Another reason why we might not want to create pure functions is when the non-deterministic nature fits within the domain of what we are building. If we are building a game of Monopoly, it is the desired effect to have a `rollDice` function return a non-deterministic result. The game of Monopoly example was no accident. Randomness is inherent in many of the games we see around us, and thus a domain where pure determinism is not the desired outcome of each function.

When we really have to panic!

When your program is in a state whereby it is impossible to continue operating normally, the typical way of handling that is by using a `panic`. While panics should be used sparingly, they are instances of the side effects that you are generating. Earlier in this chapter, we saw an example where a function would unpredictably panic during its execution. That example was artificial and a pretty bad use case for the `panic` function. But that does not mean that there are never any valid reasons to use `panic`. For example, if you are trying to reserve memory beyond the memory that is available on the system, that could be a cause for `panic`. In general, `panic` should be used to signal that the normal operation can't proceed and there's no way to gracefully continue running the application.

Two things are worth pointing out. The first one is that using the `panic` keyword should be the exception rather than the role. The second is that there is a common error handling paradigm in Go, namely returning a tuple containing a potential error value. Returning an error from a function is a distinct operation from using `panic` and serves a different use case.

How do we create pure functions?

So far in this chapter, we have taken a look at some properties of pure functions. We have also touched on some of the advantages we can gain by writing all our functions as pure functions. Now, let's look at some things we can do to make it easier to write pure functions.

Avoid global state

One of the ways we can facilitate writing pure functional code is by avoiding the global state in our programs. In Go, this comes down to avoiding the use of `const` and `var` blocks at the package level as much as possible. When you see these blocks, there's a good chance that the program state is relied upon by some functions, thus generating either side effects or having non-deterministic program execution. While it's not always possible to completely avoid such state variables, we should try to limit their use as much as possible. The way to prevent a function from relying on this state is by having the state pass to the function through a normal function argument. This is rather straightforward. The following is a small example, once using state from a `var` block and once without:

```
var (
    name = "Remi"
)
func sayHello() string {
    return fmt.Sprintf("hello %s", name)
}
func main() {
    sayHello()
}
```

We could get the same functionality as the preceding block without the `var` block by simply passing the `name` parameter as input to our function:

```
func sayHello(name string) string {
    return fmt.Sprintf("hello %s", name)
}
func main() {
    sayHello("Remi")
}
```

That's the gist of it. Next, let's look at a general way of dealing with code that contains impure elements.

Separate pure and impure functionality

As mentioned earlier, it's hard to be completely pure. We should not aim to eradicate I/O operations, API calls, and so on as by eliminating these, we'd likely be throwing out most of what makes our programs valuable. The main exercise will be in trying to create as many small, pure functions as possible and composing these into a larger program. There will still be side effects, but we'll limit their occurrence.

Bubbling up errors

A somewhat common side effect is generated by errors. Our programs end up in a state where they do not continue gracefully and there's no real way to circumvent this. One way to isolate the pure and impure aspects here is by using the Go error-handling idiom and essentially "bubbling up" errors to a common layer where they can be handled. We saw this earlier in our example of selecting random players. Since Go 1.13, there are additional built-in tools available for bubbling up errors.

Each function does exactly one thing

This is good advice in general. Generally speaking, a function should only do one thing, which significantly reduces the odds of our function creating a side effect. You find this same principle in traditional object-oriented languages as well. The industry more or less agrees that this is the way, but it's surprisingly easy to break this good intention. Take a look at the following code of a simple addition function:

```
func add(a, b int) int {
    sum := a + b
    fmt.Println(sum)
    return sum
}
```

This is not a pure function. The side effect of this snippet is that we are printing the sum value to the standard output. Sure enough, this is pretty harmless, but if our users rely on this functionality, how do we ensure this function works properly? In other words, how would you go about testing that this function prints the correct output to the screen?

A variant of this could be writing to the filesystem or a database call as part of a function where that should not be the case. Let's take a look at a function for signing up new users to a service. We expect the input to be a username and a password, and there's some logic defined on the User struct to ensure the password is compliant with password rules:

```
func createUser(username, password string) {
    u := User{username, password}
    if u.validPassword() {
        userDb.save(u)
    } else {
        panic("invalid password")
    }
}
```

The issue with this function is that it tries to do two things. First, it creates a new user struct and confirms that the password is compliant. Next, it stores the User struct in a database, assuming that the password is valid; otherwise, it panics. We could have split this into multiple functions, one for validating the password, one for storing the user, and a third function for orchestrating this:

```
func signup(username, password string) {
    user, err := createUser(username, password)
    if err != nil {
```

Example 1 – hotdog shop   71

```
            saveUser(user)
        } else {
            Panic("Could not create account")
        }
    }
    func createUser(username, password string) (User, error) {
        u := User{username, password}
        if u.validPassword() {
            return u, nil
        }
        return User{}, Errors.new("invalid password")
    }
    func saveUser(u User) {
        userDb.save(u)
    }
```

In the preceding example, we have separated the concerns, but we're still left with two impure functions. However, the problems are now more contained tough within a single function. This code is not perfect yet, and there's still room for improvement, as we will see in the next chapter. Before we go there, though, let's take a look at a more extensive example.

Example 1 – hotdog shop

For our first example, we are going to take a look at some code that has been written in an impure way, and which is pretty much violating all good sense for writing pure functions. We'll refactor this code as we go along to create more testable code, along with improving the readability and understandability of the code.

Bad hotdog shop

First, let's take a look at how not to create this hotdog shop system. We'll start by defining a constant, a global variable that dictates the price of our hotdog:

```
const (
    HOTDOG_PRICE = 4
)
```

Next, we'll create some structs. We will need a struct to represent a hotdog, as well as a struct to hold our credit card information. To keep things simple, the hotdog does not hold any state variables at the moment, while the credit card only stores the credit that is available on the card. Credit in this example is an integer value. It's not something that accurately represents a monetary value in real life, but it's good enough for this example:

```
type CreditCard struct {
    credit int
}
type Hotdog struct{}
```

With these defined, we can get to the first piece of functionality that we care about. We need a way to charge our credit card for a certain value:

```
func (c *CreditCard) charge(amount int) {
    if amount <= c.credit {
        c.credit -= amount
    } else {
        panic("no more credit")
    }
}
```

In the preceding `charge` method, we are charging the credit card for a certain amount by decreasing the credit available on the card. If there isn't enough credit available to make the charge, we are using `panic` to halt the program. For now, the main issue with this function is the use of side effects. There are two of them. First, we are using `panic` if a certain branch is hit. The next side effect is that we are changing the state of `CreditCard`. Struct immutability is a topic we'll cover in detail in the next chapter, so let's close our eyes to this issue for now and continue writing the rest of our hotdog shop. The most important function for a user is to order a hotdog. So, let's take a look at an implementation for doing so:

```
func orderHotdog(c *CreditCard) Hotdog {
    c.charge(HOTDOG_PRICE)
    return Hotdog{}
}
```

The preceding code is, again, impure code. The credit card of the user is being charged by a price defined outside of the function, using the global state. This function is doing more than one thing – it is both creating a hotdog to return to the user as well as charging their credit card.

Think about how you would test this for a second. It is possible to test this – but not convenient. You need to test or mock the credit card to also ensure a hotdog is being returned from this function. Furthermore, you have to capture a potential panic, which is not happening in the `orderHotdog` function but rather a call deeper. In addition, because `charge` is also impure, a reader of `orderHotdog` has no idea that `charge` *might* panic unless they take a look at that specific function as well. As we

Example 1 – hotdog shop   73

learned earlier, pure functional code gives us more confidence when reading code. We trust that a function does what it says it will – nothing more, nothing less. With that in mind, let's take a look at how we can refactor this code.

Better hotdog shop

In this version of the hotdog shop, we will try to address some of the issues we found in the previous example. The full code can be found at `https://github.com/PacktPublishing/Functional-Programming-in-Go./tree/main/Chapter4/Examples/HotdogShop/PureHotdogShop`.

Let's start by defining our types:

```
type CreditCard struct {
    credit int
}
type Hotdog struct {
    price int
}
type CreditError error
type PaymentFunc func(CreditCard, int) (CreditCard,
  CreditError)
```

Here, we have defined all the types we need to represent data in this small application. Our `CreditCard` struct contains an integer amount of credit, and our `hotdog` costs an integer amount as well. We have defined a **type alias** for `error` called `CreditError`, as well as a type alias for a payment function. Let's also set up some constructor-like functions for our `CreditCard` and `Hotdog`:

```
func NewCreditCard(initialCredit int) CreditCard {
    return CreditCard{credit: initialCredit}
}
func NewHotdog() Hotdog {
    return Hotdog{price: 4}
}
```

These are pretty straightforward. We will add a global variable to represent an error whereby a user does not have enough credit to perform an operation on the credit card:

```
var (
    NOT_ENOUGH_CREDIT CreditError = CreditError(errors.
      New("not enough credit"))
)
```

As you may recall, earlier, I advocated against using these types of package-level declarations. That's still true and I'd advocate avoiding using them as much as possible. For error declarations, however, this is pretty much the accepted, idiomatic way of writing Go code.

We could avoid it here and instantiate the error in-line wherever it is applicable, but that will slightly hurt the testing code, which we'll write later. In general, keep in mind that I advocate *functional programming* in Go, not *pure functional programming*.

Either way, let's write our first non-trivial function. We will rewrite the initial charge function in a pure way. The goal here is to eliminate the initial side effect that we had not by using panic but rather by returning a tuple containing a potential error:

```
func Charge(c CreditCard, amount int) (CreditCard, CreditError)
{
    if amount <= c.credit {
            c.credit -= amount
            return c, nil
    }
    return c, NOT_ENOUGH_CREDIT
}
```

As you can tell in the preceding snippet, we are not only returning an error value, but we are also returning a value of the CreditCard type. This is not the same CreditCard that is passed to the function by a caller. As we are not using a pointer to CreditCard, when a function calls Charge, **a copy** of CreditCard will be used inside the Charge function. As we are working on a copy, the c.credit -= amount statement only impacts the copy and not the original CreditCard. This is a common pitfall for newer Go programmers. In the next chapter, we will dive into immutability in greater detail and discuss the trade-off between this approach and the pointer-based function calls. But suffice it to say that this current function is *pure enough*.

This Charge function is also easily testable. Let's write a unit test to make sure the behavior is as we expect. First, we will define our test cases. The following structure is the setup for a *table-driven test*:

```
var (
    testChargeStruct = []struct {
            inputCard   CreditCard
            amount      int
            outputCard  CreditCard
            err         CreditError
    }{
            {
                    CreditCard{1000},
```

Example 1 – hotdog shop   75

```
                    500,
                    CreditCard{500},
                    nil,
            },
            {
                    CreditCard{20},
                    20,
                    CreditCard{0},
                    nil,
            },
            {
                    CreditCard{150},
                    1000,
                    CreditCard{150},    // no money is withdrawn
                    NOT_ENOUGH_CREDIT,
            // payment fails with this error
            },
        }
    )
```

In the preceding snippet, we are testing a few paths that our code can take. We can try to charge the credit card when we have more credit available than the cost, when we have the exact amount available, or when we don't have enough credit available. With this table structure, adding more test cases is trivial. Now, let's write the unit test itself, which will just run a test for each of the test cases defined previously:

```
func TestCharge(t *testing.T) {
    for _, test := range testChargeStruct {
        t.Run("", func(t *testing.T) {
            output, err := Charge(test.inputCard, test.
                amount)
            if output != test.outputCard || !errors.
              Is(err, test.err) {
                t.Errorf("expected %v but got %v\n,
                    error expected %v but got %v",
                    test.outputCard, output, test.err, err)
            }
        })
```

```
        }
    }
```

Et voilà! A complete unit test for the charge function. Something that would have been nigh impossible in the impure example. Now, let's also refactor the `OrderHotdog` function that we had earlier. As with anything, there are multiple ways to tackle this problem. The solution that we are implementing here is using a higher-order function to delay computation to a later stage. This will move the side effect of actually charging the credit card up the call chain:

```
func OrderHotdog(c CreditCard, pay PaymentFunc) (Hotdog, func()
(CreditCard, error)) {
    hotdog := NewHotdog()
    chargeFunc := func() (CreditCard, error) {
            return pay(c, hotdog.price)
    }
    return hotdog, chargeFunc
}
```

Let's unpack what's happening here. First, there's our function signature. The `OrderHotdog` function still accepts `CreditCard` as input, but also a `PaymentFunc`. Recall that we defined `PaymentFunc` as a function that takes a `CreditCard` and an `int`, and returns a `CreditCard` and a `CreditError`. The `OrderHotdog` function returns the `Hotdog` itself, as well as a function that will return a `CreditCard` and an `error`. This might be a bit confusing at first but will become clearer in the function body.

The first step is creating a new hotdog. After this, we must create a new function in-line. Recall that this is possible because Go supports functions as first-class citizens. Inside this function, we are calling `pay`, with the provided credit card, for the price of a hotdog. This is a **closure**, which we have explored in earlier chapters. Our `OrderHotdog` function then returns the hotdog and the newly created function. It's important to note that `chargeFunc` is not executed when the `OrderHotdog` function is called. No side effect is occurring in this function; the side effect is deferred to a later stage. Once again, we are going to try to isolate our side effects as much as possible. Higher up the call chain is a better place for side effects as our code is typically read from higher to lower levels of abstraction. This avoids surprises somewhere hidden in the implementation details.

With this, we have recreated the functionality of the original hotdog shop. Before we take a look at testing `OrderHotdog`, we will first look at an example of how we would use this function. In the following `main` function, we are going to order a hotdog and subsequently call the `pay` function to charge our credit card:

```
func main() {
    myCard := NewCreditCard(1000)
```

Example 1 – hotdog shop   77

```
    hotdog, creditFunc := OrderHotdog(myCard, Charge)
    fmt.Printf("%+v\n", hotdog)
    newCard, err := creditFunc()
    if err != nil {
        panic("User has no credit")
    }
    myCard = newCard
    fmt.Printf("%+v\n", myCard)
}
```

There we go – a usable example of ordering a hotdog. Let's look at how we are calling `OrderHotdog`. We are passing both the credit card, as well as the `Charge` function we wrote earlier. You can run this example in the GitHub example repository and play around with it. Let's also confirm that this code is testable by writing a unit test function.

We don't need a table-driven test for this. The `OrderHotdog` function needs to be tested to ensure that it does the following:

- Creates a new hotdog
- Creates a function that calls the payment function
- Returns both the hotdog and the function

Our test function will confirm that a new hotdog has been created and that a payment function is called. As this is a unit test, we don't care about the payment function itself. We will mock a payment function to ensure that it is called from the returned function. The actual `charge` function is tested separately, as we saw earlier:

```
func TestOrderHotdog(t *testing.T) {
    testCC := CreditCard{1000}
    calledInnerFunction := false
    mockPayment := func(c CreditCard, input int) (CreditCard,
      CreditError) {
        calledInnerFunction = true
        testCC.credit -= input
        return testCC, nil
    }

    hotdog, resultF := OrderHotdog(testCC, mockPayment)
    if hotdog != NewHotdog() {
```

```
            t.Errorf("expected %v but got %v\n", NewHotdog(),
                hotdog)
    }


    _, err := resultF()
    if err != nil {
            t.Errorf("encountered %v but expected no error\n",
                err)
    }
    if calledInnerFunction == false {
            t.Errorf("Inner function did not get called\n")
    }
}
```

In the preceding code, we are strictly testing that our function is creating the correct values for the hotdog and the closure. A correct closure function in this case implies that the function that is returned calls the payment function that is passed to it. Notice how we could mock away the original behavior and create a bool to ensure that the function is called. Once again, this is the power of having first-class functions in Go.

Summary

In this chapter, we looked at pure functional programming. First, we looked at what exactly it means for a programming language to be pure and functional as opposed to impure and functional. Next, we took a more detailed look at how pure code can help improve testability by eliminating side effects. We also learned that pure code gives readers more confidence in the code that they are reading since functions are more predictable and won't change the state of the system. We also discussed when we should not use pure functions, such as when dealing with functions that should generate random behavior for games or functions that deal with I/O.

Although we have only briefly touched on it, we have seen how immutability plays a core part in writing pure functions by not changing the values of structs. In the next chapter, we will take a deep dive into immutability, how it does (or doesn't) impact performance, and how we can leverage this in combination with pure functions to write more maintainable code.

5
Immutability

In this chapter, we will look at immutability. We are going to cover what exactly it means to be immutable, and how the Go language helps preserve immutability at the struct-level. To understand how this works, we will take a look at how Go handles pointers and references to objects, what the performance implications are, and how to decide between the pointer-reference trade-offs. We will also dive into the implications of garbage collection, unit testing, and *pure* functional programming.

These are the main topics that we will cover in this chapter:

- What is immutability?
- How to write immutable code
- How do pointers and references work in Go?
- Analyzing the performance of mutable and immutable code
- Examples of concurrency and testing with immutable code

Technical requirements

For this chapter, you can use any Go version at or above Go 1.18 as we will be using generics in some of the later examples. You can find all the code on GitHub at `https://github.com/PacktPublishing/Functional-Programming-in-Go./tree/main/Chapter5`.

What is immutability?

When we talk about immutability in this chapter, we are talking about structs that have a state that does not change over time. In other words, when a struct is created, that is how that specific struct will be represented during its lifetime. We can still create new structs and delete old ones. So, the state at the system level will effectively change by new structs being created and old structs being deleted. This has several advantages:

- First, because our structs are not changing, we can safely pass data to a function and know that, whatever happens, the copy that we passed to the function will remain intact.

- Secondly, immutable structs make it easier to write correct, concurrent code. As the state of the struct cannot be changed by any function calling it, we can safely parallelize execution and call multiple functions using the same struct as input data.

- And third, this makes our code easier to reason about. At each step of the way, the state of our struct is more predictable.

Immutability is not just something that we strive for when writing functional code. In many object-oriented programming languages, it is preferred to write immutable code. The reason it deserves mention in this book is that it ties in nicely with pure functions, which we saw in the previous chapter. If you want to write true *pure* functional code, you need immutable structs. If you make a change to a struct in a function, that would count as having a side effect. Recall from the previous chapter that we will try to eliminate side effects as much as possible. That said, almost everything in this chapter can still be applied to traditional object-oriented languages as well.

Immutability at the data layer

Immutability is a powerful concept that we can apply to the programs we write. But it also appears as a concept for the data that we store. If we are writing software that deals with extremely sensitive data, such as an **Electronic Health Record** (**EHR**), we likely want the data to be immutable. That is to say, whenever some information in our EHR changes, we want this change to be completely traceable. That way, the entire history of your EHR is visible at any time.

By having the medical data immutable, you can always look at what the file looked like in the past. For example, you can look at any blood test the patient has done or any notes that were previously taken. It also helps to serve as an auditable log – each change to the record is traceable. Imagine that a doctor accidentally deletes the result of a blood test. If your data storage is immutable, the blood test will not be deleted at the data layer (but rather marked as "deleted" so that the application layer can choose not to display it to a user). It also protects against ill intent – if a bad actor gained access to the application and decided to start changing the text of the doctors' notes, this would show up as *new* notes. The original notes would still be there, at least in the data layer.

Imagine what would happen if we did not have immutability, and the actual information updated each time new data became available. This would be less than ideal. Imagine that each blood test overrides the past results – this would obfuscate any trends in your medical history, erasing valuable information to medical practitioners. Or worse, once a medical image is deleted, it would stay deleted, and the patient would have to undergo the same series of tests. Not only is this bad for the patient's experience, but in some countries it is also costly.

This idea of traceability and immutability at the data layer, in some ways, culminated in what is now called **the blockchain**. While I don't know of any mainstream EHR systems implemented on top of blockchain databases, there are at least some efforts being made by companies around the world to make this a reality. And it would make some sense to do so.

Blockchain databases are immutable by default. Apart from being suitable for the EHR examples mentioned previously, it is currently being used for trading currency. In a blockchain database, the entire history of the block is visible. When an update is made to a block, a new block is added to the chain with the updated information, rather than it overriding the existing block. This is how cryptocurrencies can model financial transactions. There is more depth to it than what I've explained here, as I've omitted a concrete explanation of how a blockchain can guarantee immutability and provide tamper mechanisms.

A deep dive into immutability at the data layer is beyond the scope of this book, but hopefully, this short overview served as a good starting point to explore these ideas further.

How to write immutable code in Go

When we talk about immutability in Go, we are specifically focusing on how to have immutable structs in our code. At the core of this, we have to take a look at how Go uses pointers and the difference between pass-by-value and pass-by-reference. This is something that trips up new Go programmers, and there is a sufficient amount of edge cases where even more seasoned Go programmers will occasionally shoot themselves in the foot.

In essence, it comes down to whether or not we are using pointers in our code when passing around structs to functions. If our code is entirely free of pointers, then we would also be writing immutable code.

To demonstrate this, take a look at the following piece of code. We have a struct to define a person, and a function to change the name of this person:

```
type Person struct {
    name string
    age  int
}
func main() {
```

```
    p := Person{
        name: "Benny",
        age:  55,
    }
    setName(p, "Bjorn")
    fmt.Println(p.name)
}
func setName(p Person, name string) {
    p.name = name
}
```

The outcome of this function, perhaps contrary to expectation, is Benny. The setName function has not changed the name of the Person object. Eventually, we all get used to the idea that to update structs in a function, we need to use a pointer instead:

```
func main() {
    p := Person{
        name: "Benny",
        age:  55,
    }
    setName(&p, "Bjorn")
    fmt.Println(p.name)
}
func setName(p *Person, name string) {
    p.name = name
}
```

Now, when we run this code, the output is Bjorn, as we expected. The difference between these two examples is that in the first example, we are using pass-by-value, while in the second, we are using pass-by-reference.

If we look at what is happening in the first function, we will see that our Person object is being copied and that this copy is then passed to the setName function. Thus, every operation that we do on this struct is happening on the copy itself, and not on the actual object. However, in the second example, by using a pointer, we have access to the actual Person object and not just a copy. Under the hood, the second example passes an address (pointer) to the struct. The syntax of Go obfuscates some of the pointer referencing and dereferencing for us, which makes it seem like a rather small change.

In general, we want to keep our code immutable. Hence, we want to avoid using pointers in our code. How, then, do we update our structs? The `setName` function provides useful functionality to our system. Recall that although we cannot change the state of the objects we are using, we are still free to create and destroy them. The solution is to create a new object that has all the properties of our original object, with some changes applied. To continue our previous example, let's refactor the `setName` function to achieve the desired functionality:

```go
func main() {
    p := Person{
        name: "Benny",
        age:  55,
    }
    p = setName(p, "Bjorn")
    fmt.Println(p.name)
}
func setName(p Person, name string) Person {
    p.name = name
    return p
}
```

In the preceding example, you can see the core change in which we need to update structs without breaking our immutability concern. We achieve this by having functions accept copies (pass-by-value) as input and return a new struct with the changes applied. In our calling function, we now have the choice of whether or not to keep both objects or discard the original and keep only the newly returned object.

This syntax should be quite familiar to Go programmers, as this is similar to what we do when working with slices. For example, if we wanted to add a value to a slice, we would write code like the following:

```go
func main() {
    names := []string{"Miranda", "Paula"}
    names = append(names, "Yvonne")
    fmt.Printf("%v\n", names)
}
```

This code would return `[Miranda Paula Yvonne]`. When working with immutable structs, our syntax will look similar to this.

Writing immutable code for collection data types

Earlier, we saw how easy it is to change functions from immutable to mutable. We simply replace a function that takes a pointer with a function that accepts a value and returns a new value. The story changes a bit when working with the collection Map data type, as becomes apparent in the following example:

```
func main() {
    m := map[string]int{}
    addValue(m, "red", 10)
    fmt.Printf("%v\n", m)
}
func addValue(m map[string]int, colour string, value int) {
    m[colour] = value
}
```

The output of this code is [red 10]. Although we are not using a pointer in the addValue function, the function is not operating on a copy of the map but is operating on the map itself. **Maps always act like pass-by-reference in Go.**

If we try a similar setup with slices, another collection data type, it works as expected:

```
func main() {
    names := []string{"Miranda"}
    addValue(names, "Yvonne")
    fmt.Printf("%v\n", names)
}
func addValue(s []string, name string) {
    s = append(s, name)
}
```

The output here is Miranda. Using pointers, we can once again make the function mutable:

```
func main() {
    names := []string{"Miranda"}
    addValue(&names, "Yvonne")
    fmt.Printf("%v\n", names)
}
func addValue(s *[]string, name string) {
    *s = append(*s, name)
}
```

If we run the preceding code, the output will be [Miranda Yvonne]. It is common enough in Go that seasoned programmers are used to this, but it can trip up the more novice Go programmer.

Measuring performance in mutable and immutable code

A common complaint about immutable code is that it is less performant than its mutable counterpart. Even without doing a deep dive into the performance characteristics of the Go runtime, this seems like a reasonable statement. After all, in the immutable variant, a new copy of an object is spawned for each function call. In practice, however, these differences in performance are often negligible.

Still, even if there would be a significant performance impact, you need to question if the performance sacrifices make sense in your context. In return for some performance, you are getting thread-safe, easy-to-maintain, understand, and test code. As engineers, it is often extremely tempting to go for the most optimal solution, using as little memory and CPU time as possible. However, for many real-world applications, the performance impact is small enough that this is not something the end user would notice. And for other engineers maintaining your code, they'd often want something more understandable rather than something faster.

Unlike other languages, Go will take somewhat of a performance hit due to being garbage collected. If you want to squeeze every ounce of performance out of your system, perhaps Go is not the right tool for the job either. With this out of the way, we should take a look at actual benchmarks and dive a bit deeper into the performance implications of immutable code.

Benchmarking functions

While we can reason about the performance of a function in abstract terms, such as space-time complexity, to get a real sense of performance, we should do performance testing. After all, the runtime complexity of mutable and immutable functions can be quasi-identical. Caring about the implementation of pointers is too low-level to be considered. So, for that reason, we will set up a test to determine which performance is worse. As a reminder, the assumption here is that the mutable code, using pointers, will be faster than our immutable variant. The underlying reason for that assumption is that copying structs is a more costly operation than passing pointers to a function.

Let's set up two constructor-like functions, one for an immutable version and the other for a mutable version. The first function creates a Person object, and then passes that function to a function to set a name for the person, and subsequently to another function that sets an age for the person:

```
func immutableCreatePerson() Person {
    p := Person{}
    p = immutableSetName(p, "Sean")
    p = immutableSetAge(p, 29)
    return p
}
```

```
func immutableSetName(p Person, name string) Person {
    p.name = name
    return p
}
func immutableSetAge(p Person, age int) Person {
    p.age = age
    return p
}
```

Here, we can see that the `Person` object first gets copied to `immutableSetName` and afterward is copied again to `immutableSetAge`. Finally, we return this `Person` to the calling function.

Now, let's also set up a mutable version of this code. In the mutable version, we are creating a `Person` object. But, when passing it to the mutable functions to set a name and an age, we are going to pass a pointer to our object instead:

```
func mutableCreatePerson() *Person {
    p := &Person{}
    mutableSetName(p, "Tom")
    mutableSetAge(p, 31)
    return p
}
func mutableSetName(p *Person, name string) {
    p.name = name
}
func mutableSetAge(p *Person, age int) {
    p.age = age
}
```

Here, we can see that pointers are used to avoid copying the `Person` object between functions. One thing to point out in these examples is that these two functions are identical in Go:

```
func mutableSetName(p *Person, name string)
```

And with the function bound to an object instead:

```
func (p *Person) mutableSetName(name string)
```

There are some practical differences in how we call these functions and implications for function name collisions. That said, their performance characteristics are identical for both the mutable and immutable examples.

With that out of the way, let's write our benchmark. Go has built-in benchmarking support, just like it has built-in testing support. That makes our job of writing benchmarks rather easy since the entire code to benchmark fits on a single page:

```
package pkg
import "testing"
func BenchmarkImmutablePerson(b *testing.B) {
    for n := 0; n < b.N; n++ {
        immutableCreatePerson()
    }
}
func BenchmarkMutablePerson(b *testing.B) {
    for n := 0; n < b.N; n++ {
        mutableCreatePerson()
    }
}
```

With this built-in benchmarking support, we can run our benchmark with the following command:

```
go test -bench=.
```

Averaged out over a couple of runs, on my **Amazon Web Service** (**AWS**) EC2 instance, I get the following result:

```
BenchmarkImmutablePerson          0.3758 ns/op
BenchmarkMutablePerson            0.3775 ns/op
```

The concrete values for these ns/op properties are going to be different on your machine, so don't focus too much on the concrete values. What should be surprising here is that our immutable code outperforms our mutable code.

To understand what's going on, we need to take a look at garbage collection, and stack versus heap allocation.

Understanding stacks, heaps, and garbage collection

Garbage collection is a topic that is complex enough that it probably deserves a full chapter in its entirety. We will take some shortcuts here and look at a sufficiently deep understanding of this process but will simplify some steps. Go itself is open source and has good documentation.

Reclaiming memory through garbage collection

Go is a garbage-collected language, which means that memory management is taken care of by the Go runtime. This reduces the effort required on the programmer's side as it takes away from the need to manually manage memory. This can eliminate or reduce the odds of having certain types of bugs in your code, such as memory leaks.

With automatic garbage collection, we, the programmers, don't have to think about managing the memory of our application. Memory will be reserved for us, and later given back to the system, without our intervention. To make this work, the Go runtime needs to do some behind-the-scenes work. Essentially, the runtime will trigger a "garbage collection" process to free up memory. It does so by temporarily freezing our application, checking which objects are no longer required, and removing them from the working memory of our application. There are different ways of figuring out which objects are no longer required, and some mechanisms to delete them throughout our program's lifetime. Typically, a garbage collector will try to figure out if there are still any references to a piece of data. If there's a reference to the data, it is still accessible by your program and thus should not be deleted.

To understand how this process impacts performance, it helps to think of garbage collection as a *stop-the-world* process. This means it completely stops all execution, identifies garbage, and removes it to free up memory. In practice, Go uses multiple threads to identify the garbage objects. This approach is called the *concurrent mark-and-sweep garbage collector*. Regardless of this being concurrent, there is still performance overhead. When people are deciding on which language to use for their application, the garbage collection overhead pops up surprisingly often in conversation. This is most apparent when the decision has to be made between Go, C/C++, or Rust.

While the performance impact of garbage collection has been reduced in the more recent Go version, the impact cannot be completely erased. There are ways of tweaking the garbage collector's behavior in Go, but in general, that would not be a recommended approach. Often, a suboptimal implementation of an algorithm would outweigh the negative impact felt by garbage collection.

Stacks and heaps

Our next topics to discuss are stacks and heaps. There are two types of memory available at runtime, namely a stack and a heap. A stack is a **Last-In, First-Out** (**LIFO**) data structure. This means that when data is removed from the stack, the last item to have been inserted will be deleted. Go uses a stack to store data in a chain of function calls, this includes local variables, the function's input parameters, and more.

When a function is called, the data of this function is pushed to the top of the stack. When the function is done executing, this data is removed from the stack. Thus, the stack is continuously growing and shrinking while functions are being called in your application. There is a limited amount of space available for the stack; exceeding this leads to an error well known as a *stack overflow*. The elements on the stack can be thought of as having a limited lifetime as they are removed from memory quickly (at the end of a function).

The heap, on the other hand, is shared memory for the lifetime of your application. The data that is stored here is not limited to the lifetime of a function. This means that this data can be referenced (pointed to) from multiple places in your application. To avoid the heap from continuously expanding, the heap memory is managed by the garbage collector. The garbage collector will scan the memory in the heap to figure out if it's still needed or not. If the data is no longer needed, it is deleted.

In the stack and heap implementation, it is cheaper to reclaim memory from a stack than it is from a heap. The stack does not need a garbage collector that "stops the world" to scan for objects to delete. Thus, if we could allocate as much as possible on the stack instead of the heap, our programs would run faster. This is not always possible as there is data that we want to keep alive outside of the context of a single function. In addition to this, heap allocations tend to be slower than stack allocations, as the memory required for a heap allocation needs to be reclaimed from a memory pool – a set of memory that Go has claimed from the operating system. This is a potentially slow operation as your program waits for the memory to become available.

To understand how this impacts the performance of the immutable and mutable example we looked at previously, we need to understand how Go chooses where to store a variable. In theory, this sounds simple – if data is needed only in a single function it is a stack variable; otherwise, we have to store it on the heap. In practice, though, there are a few more things to consider.

First, the compiler will try to prove that a variable is local to a single function. The compiler does this through a process called *escape analysis*, where it looks for variables that escape the context of a single function. If a variable is not local to a single function, it stores it on the heap. Another piece of information the Go runtime will look at is the size of the data. It makes more sense to store large data on the heap rather than on the stack, as the stack is typically more limited in space. Stack space is a real issue that we will explore in some more depth when we discuss recursion in the next chapter.

How does this tie into our conversations of mutability with pointers? In the example code, which we used to benchmark the two functions, the immutable code can allocate all memory on the stack. The mutable example is not so lucky and will allocate data on the heap since we are using pointers, which is the context that escapes a single function. Thus, the impact we are seeing in performance is caused by the garbage collector reclaiming memory.

It's important to note that the concrete implementation of the garbage collector, and even the algorithms for escape analysis, can and do change over time. To understand how garbage collection works in the latest version of Go, it's best to read the documentation of that version.

Seeing escape analysis in action

Let's explore the behavior of escape analysis in Go to show that our reasoning makes sense. First, we will change our code slightly by adding a pragma to avoid the compiler from in-lining our function. A pragma is a special comment in Go that gives some instructions to the compiler. We will add this to each function so that they will all have the comment present, as shown here:

```
//go:noinline
func immutableCreatePerson() Person {
```

```
    p := Person{}
    p = immutableSetName(p, "Sean")
    p = immutableSetAge(p, 29)
    return p
}
```

This means the functions are not erased by the compiler. Function inlining is a compiler optimization process that occurs behind the scenes to speed up the execution of our programs. Once again, this deserves a chapter on its own but is outside the scope of this book.

Once we have added the pragma to each function, we can build our application with the following command:

```
go build -gcflags '-m -l'
```

This tells the Go compiler to explain to us where escape analysis decisions are being made, and what the outcome of these decisions is. When we look at the output, we get this:

```
# github.com/PacktPublishing/Chapter5/Benchmark/pkg
./person.go:17:23: leaking param: p to result ~r0 level=0
./person.go:17:33: leaking param: name to result ~r0
  level=0
./person.go:23:22: leaking param: p to result ~r0 level=0
./person.go:37:21: p does not escape
./person.go:37:32: leaking param: name
./person.go:42:20: p does not escape
./person.go:30:7: &Person{} escapes to heap
```

What this shows us is that, on line 30, our Person is escaping to the heap. And when an object escapes to the heap, this eventually has to be picked up by the garbage collector so that our memory space can be reclaimed.

Many things are happening behind the scenes, and we simplified some of how garbage collection works in Go. But overall, this should serve as an example of why the assumption that pointers and mutable code are faster than immutable code without pointers is invalid.

When to write mutable functions

So far, this chapter has indexed heavily on why we prefer to write immutable functions. But there are some instances in which it makes sense to write mutable functions either way. The only real reason is performance. As we saw earlier, the performance implications can often be ignored, but not always. If you are using structs that contain a lot of data, copying that over to each function can negatively impact the performance sufficiently to cripple your application. The only real way of knowing whether this is the case is by adding performance metrics to your application. Even so, a trade-off must be made between more performant code and more maintainable code. Oftentimes, trying to squeeze more performance out of your application hinders long-term maintainability.

Another possible reason to write mutable code using pointers is for resources that need to be singularly unique within your application. If you're implementing traditional object-oriented patterns in your code, you might have implemented the singleton pattern. If you want to have a true singleton, you should be using a pointer rather than copying over the singleton. Otherwise, you will have multiple copies of your singleton available in different functions, potentially each with a different state. Whether or not having singletons in your code is a good idea is a discussion for a different book.

What are functors and monads?

In the previous chapter, we discussed the concept of function purity. A function should not produce any side effects and should be idempotent. In this chapter, we have seen how structs can be made immutable, and how this ties into function purity. As mentioned earlier, even in a purely functional language, in which side effects are eliminated as much as possible, you still have desirable side effect behavior. For example, getting input from a user, or writing data to a database, are both side effects that add value to a program.

In this section, we will try to build an understanding of how pure functional languages can achieve this. We'll also look at an implementation in Go to achieve the same results, building on top of our knowledge about immutable structs and pure functions.

To preface this section, it is commonly said that there are too many monad explanations already and all of them are wrong or lacking in some manner. There are many books about functional programming, or blog posts and videos, that try to offer a good explanation. The fact that new explanations are being offered so frequently should give you an idea as to the complexity of the topic. I don't have the lofty goal of offering the "final monad explanation that is ever needed." Rather, I will try to cut it down to the core idea and keep it as close to what's practical as possible. As such, we will stay away from the deeper theoretical layers of category theory. What follows is, hopefully, a *good enough* explanation of the idea rather than a perfectly holistic explanation.

What's a functor?

Before we can demonstrate what a monad is, we need to understand what a functor is. A functor, simply put, is a function that can apply an operation to each element contained in a data structure. In Haskell, the implementation of this function is called `fmap`. In Go, this function might look something like this:

```
func fmap[A, B any](mapFunc func(A) B, sliceA []A) []B
```

In the preceding type signature, we are using slices. A slice is a data type that contains other data elements. The `fmap` implementation does not have to operate on slices – any data structure that holds data elements will do, such as pointers (they optionally hold a data element), functions themselves, trees, or as we will see in the next few pages, a monad.

If we were to write an implementation of `fmap` in Go to operate on slices, as shown in the function signature previously, we would simply call the provided `mapFunc` for each element in `sliceA`. The result of this would be stored in the new slice, `sliceB`:

```go
func fmap[A, B any](mapFunc func(A) B, sliceA []A) []B {
    sliceB := make([]B, len(sliceA))
    for i, a := range sliceA {
        sliceB[i] = mapFunc(a)
    }
    return sliceB
}
```

Notice the use of generics in the preceding example, which we can use to map between two `any` types. But the input is A and the output is B. The map function thus **changes the type** of our data.

Let's take a look at how we would use this function. Imagine that we have a slice of integers, and we want to transform this into a slice of strings. We could use our `fmap` function to do exactly this. All we need to do is provide `fmap` with a function that takes an integer and returns a string:

```go
import (
    "fmt"
    "strconv"
)
func main() {
    integers := []int{1, 2, 3}
    strings := fmap(strconv.Itoa, integers)
    fmt.Printf("%T transformed to %T - %v\n", integers,
      strings, strings)
}
```

When we run the preceding function, we get the following output (recall that `%T` prints the type of the variable):

```
[]int transformed to []string - [1 2 3]
```

This tells us that our int, `slice`, was transformed into a string slice, and the values contained are, to no surprise, [1, 2, 3].

This is pretty much what a functor is. It's a function that transforms all data in a given data structure into data of a different type. The fmap implementation is a pure, higher-order function.

From functor to monad

The next step is getting from a functor to a monad. So, what exactly is a monad? When we aim for a somewhat theoretical description of the monad, we might get something such as the following.

A monad is a software design pattern. It is a data type that can combine functions of similar types and wrap the results of a non-monad type into a new monadic type offering additional functions. For a type to be a monad, it needs to have two functions defined:

1. **A function to wrap a value of the T type into Monad[T]**

2. **A function to combine the function of the Monad[T] type**

We will demonstrate the monad with a practical example. A **monad** type is a *container* that has an underlying concrete type (for example, String). A popular monad is the Maybe monad, also known as Optional in some programming languages. The Maybe monad is a type that *potentially* contains a concrete value, but also might be empty.

To model the Maybe monad in Go, we will use an interface that defines the operations on our struct. Next, we will also create two implementations, one for when a value is present, and one for when the value is absent:

```
type Maybe[A any] interface {
    Get() (A)
    GetOrElse(def A) A
}
```

In the preceding interface implementation, we have defined two functions: Get and GetOrElse. More can be defined; the concrete functions don't matter as much. What's important is that we have a way to model values that might or might not be present.

Notice that we are not using pointers here, we're only using concrete types. The Maybe monad is often introduced to avoid pointers. By avoiding pointers, we can eliminate a class of errors that happen at runtime when functions are called on *null pointers*. The null, or nil in Go, also does not make real sense from a type taxonomy perspective. The nil pointer belongs to every type, meaning there's no real useful information in there, and we want our type system to be as declarative as possible. (Go does have a typed nil, on which functions can be called safely. Still, exercise caution whenever using this. It's not common behavior in programming languages and can trip up even seasoned Go programmers.)

George Hoar, who first introduced the null pointer concept, called this his "*billion-dollar mistake.*"

The two implementations that we will use for modeling the presence and absence of a value are `Just` and `Nothing`, respectively. These names have been borrowed from Haskell; you'll find different names for these values in different programming languages. `Just` signals a concrete value is present, while `Nothing` signals the absence thereof. We will start by implementing the value present use case, with the `JustMaybe` type:

```
type JustMaybe[A any] struct {
    value A
}
func (j JustMaybe[A]) Get() (A) {
    return j.value
}
func (j JustMaybe[A]) GetOrElse(def A) A {
    return j.value
}
```

The preceding code adheres to the `Maybe` interface. As such, we can use a `JustMaybe` as an instance of `Maybe`. To implement the absence of a value, we'll implement the analogous `NothingMaybe`:

```
type NothingMaybe[A any] struct{}
func Nothing[A any]() Maybe[A] {
    return NothingMaybe[A]{}
}
func (n NothingMaybe[A]) Get() (A) {
    return *new(A)
}
func (n NothingMaybe[A]) GetOrElse(def A) A {
    return def
}
```

The implementations are rather straightforward for each function. Perhaps the most surprising thing is the `return` statement in `Get` for a `NothingMonad`, where we wrote:

```
return *new(A)
```

This statement returns a new instance of A, but A is an unknown value at compile time. By using new, we can instance it, but it'll return a pointer value, which we will dereference to return a concrete value.

Next, let's also create constructor-like functions for these two implementations, which are functions that can wrap a value of a given type into the monadic representation. Recall that this is a requirement for our monad pattern:

```
func Just[A any](a A) JustMaybe[A] {
    return JustMaybe[A]{value: a}
}
func Nothing[A any]() Maybe[A] {
    return NothingMaybe[A]{}
}
```

These two implementations will let us implement both the presence and absence of a given value. For example, we could now use these in a function:

```
func getFromMap(m map[string]int, key string) Maybe[int] {
    if value, ok := m[key]; ok {
        return Just[int](value)
    } else {
        return Nothing[int]()
    }
}
```

In the preceding function, we are getting a value from a map by looking up a given key. If a value is present, we return the `JustMaybe` implementation of our monad; otherwise, we return the `NothingMaybe` implementation.

Convenience functions can be written, such as `fromNullable(*value)`, which would return either a `JustMaybe` or a `NothingMaybe` by checking if the value passed to the function is present.

Remember that our monad type is a data structure that holds underlying elements. As such, we can implement the `fmap` function on this type as well. In this implementation, we will turn a `Maybe` of type A into a `Maybe` of type B. We need to provide a function to map from the underlying type A to the underlying type B to accomplish this:

```
func fmap[A, B any](m Maybe[A], mapFunc func(A) B) Maybe[B]
{
    switch m.(type) {
    case JustMaybe[A]:
        j := m.(JustMaybe[A])
        return JustMaybe[B]{
            value: mapFunc(j.value),
```

```
        }
    case NothingMaybe[A]:
        return NothingMaybe[B]{}
    default:
        panic("unknown type")
    }
}
```

In the preceding code, we are using a type switch to determine what type our Maybe monad is to figure out if it represents the JustMaybe or NothingMaybe implementation. If the type matches JustMaybe, we will map the underlying value from type A to type B, and return this wrapped in a new monad.

This is an incomplete definition of a monad, but a practical implementation of one such instance. This concept can be pushed further, but Go does not provide a convenient way of exploring this further, so it would not often be used in the real world.

Summary

In this chapter, we touched upon immutability in Go. We took a small refresher on how immutability works in Go, by either pass-by-value or pass-by-reference. We learned that pointers do not guarantee that your code will be more performant than if you avoid them. We also discussed some of the benefits of immutable code, such as improving the readability and understandability of the code base. We also touched on how this makes concurrency easier to implement correctly, as the state is not mutated between functions.

Finally, we wrapped up the discussion of pure functions that we started in the previous chapter by looking at monads and a practical implementation thereof with the Maybe monad.

In the next chapter, we will explore some must-have functions for writing code functionally.

Part 2:
Using Functional
Programming Techniques

After we have established the basic ideas of functional programming and see how they relate to the object-oriented paradigm, we will move on to this part. Here, we will look at how functional programming can be leveraged to compose larger programs while still at the class level. We'll learn about solving problems iteratively versus recursively, the three important categories of function types, and how to chain functions together for more readable code.

This part has the following chapters:

- *Chapter 6, Three Common Categories of Functions*
- *Chapter 7, Recursion*
- *Chapter 8, Readable Function Composition with Fluent Programming*

6

Three Common Categories of Functions

In the preceding chapters, we have looked at some of the core components of functional programming. We have discussed how to write functions that adhere to both functional programming and pure functional programming.

In this chapter, we are going to look at some practical implementations of functions that leverage these concepts. These are the categories and topics we will cover:

- The first category we will look at is predicate-based functions
- Then, we will take a look at data transformation functions, which maintain the structure of our data (more on that later)
- Finally, we will take a look at functions, which transform the data and reduce the information into a single value

This is not meant to be an exhaustive list, but with these three categories, we can build a large portion of our day-to-day applications.

Technical requirements

For this chapter, you can use any Go version at or above Go 1.18, as we will be using generics in some of the later examples. You can find all the code on GitHub at `https://github.com/PacktPublishing/Functional-Programming-in-Go./tree/main/Chapter6`.

Predicate-based functions

The first type of functions that we will explore is predicate-based functions. A **predicate** is a statement that can be evaluated as either true or false. Typically, in a language without a higher-order function, this would be achieved by using `if` statements inside the body of a function. A common use case is to filter a set of data into a subset that matches a specific condition – for example, given a list of people, return all of those who are older than 18 years old.

To start, we can introduce a type alias for a function that defines the type signature of a predicate:

```
type Predicate[A any] func(A) bool
```

This type alias tells us that the function takes an input with a type of `A`, which can represent the `any` type in our program, but needs to return a `bool` value. This type uses generics, which were introduced in Go 1.18. We can now use this type in every place at which a predicate is expected. The first function that works using predicates is the simple `Filter` function.

Implementing a Filter function

The `Filter` function is a staple within the functional programmer's toolbox. Let's imagine that we don't have higher-order functions available, and we want to write a `Filter`-like function. For this, let's assume that we have a slice of numbers, and we want to filter all the numbers that are larger than 10. We could write something such as this:

```
func Filter(numbers []int) []int {
    out := []int{}
    for _, num := range numbers {
        if num > 10 {
            out = append(out, num)
        }
    }
    return out
}
```

This works well enough, but it's not flexible. In this case, this function will always just filter for numbers larger than 10. We could make it a bit more flexible by adjusting the threshold value using an input parameter for our function. With a trivial change, we get the following function:

```
func Filter(numbers []int, threshold int) []int {
    out := []int{}
    for _, num := range numbers {
        if num > threshold {
```

```
                    out = append(out, num)
            }
      }
      return out
}
```

This gives us a more flexible `Filter` function. However, as we all know, requirements change often, and users need new functionalities on an existing system almost ad infinitum. The next requirement for our function is to optionally filter for either *larger than*, or, in some cases, *smaller than*. Thinking about this for some time, you might realize that this could be implemented as two functions (the function body is omitted in snippets, as it's a trivial change):

```
func FilterLargerThan(numbers []int, threshold int) []int {
..
}
func FilterSmallerThan(numbers []int, threshold int) []int {
..
}
```

Sure enough, this would work – but the work never stops. Next, you have to implement a function that can filter for numbers larger than a given value but smaller than another. Then, our users become really into odd numbers, so there needs to be a filter for finding all odd numbers as well. Later on, the user asks you to count the exact amount of times a certain value appears, so you also need a filter for a certain value *exactly* in your list of numbers. You get the point; we can create a bunch of functions that suit all these use cases, but that approach does not sound like the best option.

One of the benefits of having a language with support for higher-order functions is that we can reduce repetitive implementations and abstract our algorithm. All of the aforementioned use cases fit within a function often called `Filter` in functional programming languages. The implementation of the `Filter` function is rather straightforward. The basic operation it supports is to iterate over a container, such as a *slice*, and apply a predicate function to every data element contained within the container. If the predicate function returns `true`, we will append this data element to our output. If not, we simply discard elements that did not match.

As we want to follow the best practices for implementing these functions, these functions will be pure and immutable. The original slice will never be modified within our filter functions, and neither will the elements contained therein:

```
func Filter[A any](input []A, pred Predicate[A]) []A {
      output := []A{}
      for _, element := range input {
            if pred(element) {
```

```
                    output = append(output, element)
            }
        }
    return output
}
```

This `Filter` implementation is a pretty typical implementation that you will find in many functional (and multi-paradigm) programming languages. Using higher-order functions in this way, we can essentially make part of an algorithm configurable. In other words, we abstract our algorithm. With the `Filter` function, the actual predicate part of an `if` statement is customizable.

Notice that we have implemented this using *generics*. `Filter` does not care what data types it is working with. Anything that can be stored in a slice can be passed to the `Filter` function. Let's look at how we would use this in practice by creating some of the functions we discussed earlier. We will start off by implementing `LargerThan` and `SmallerThan` filters:

```
func main() {
    input := []int{1, 1, 3, 5, 8, 13, 21, 34, 55}
    larger20 :=
        Filter(input, func(i int) bool { return i > 20 })
    smaller20 :=
        Filter(input, func(i int) bool { return i < 20 })
    fmt.Printf("%v\n%v\n", larger20, smaller20)
}
```

The functions that we are passing to `Filter` as input are a tad verbose, as at the time of writing, Go does not have syntactic sugar for creating anonymous functions. Notice how we did not have to duplicate the body of our `Filter` function for this implementation.

Implementing other filters, such as *larger than X but smaller than Y* or *filter even numbers*, are equally easy to implement. Remember that we only have to pass the `if` statement logic each time and the iteration of the list is taken care of by the `Filter` function itself:

```
func main() {
    input := []int{1, 1, 3, 5, 8, 13, 21, 34, 55}
    larger10smaller20 := Filter(input, func(i int) bool {
        return i > 10 && i < 20
    })
    evenNumbers := Filter(input, func(i int) bool {
        return i%2 == 0
    })
```

```
    fmt.Printf("%v\n%v\n", larger10smaller20, evenNumbers)
}
```

By implementing this with generics, our `Filter` function can work with any data type. Let's see how this function would work with the `Dog` struct that we have used in earlier chapters.

Recall that our struct for `Dog` had three fields: `Name`, `Breed`, and `Gender`:

```
type Dog struct {
    Name    Name
    Breed   Breed
    Gender  Gender
}
```

This snippet omits the `const` declarations for `Breed` and `Gender`, as well as the type aliases. These are the same as those in *Chapter 3*, and the full implementation can be found on GitHub: `https://github.com/PacktPublishing/Functional-Programming-in-Go./tree/main/Chapter3`.

Because we have used generics in the implementation of our `Filter` function, this will work on any data type, including custom structs. As such, we can use the function as is without any changes. Let's implement a filter for all dogs that are of the `Havanese` breed:

```
func main() {
    dogs := []Dog{
            Dog{"Bucky", Havanese, Male},
            Dog{"Tipsy", Poodle, Female},
    }
    result := Filter(dogs, func(d Dog) bool {
            return d.Breed == Havanese
    })
    fmt.Printf("%v\n", result)
}
```

That's all there is to it. Next, let's look at some other functions that use predicates.

Any or all

It is common to have to make sure that either *some* elements or *all* elements match a certain condition. The use case for abstracting this into a higher-order function is the same as for the `Filter` function. If we do not abstract this, a separate `All` and `Any` function would have to be implemented for each use case. While these are not found as often in multi-paradigm languages or object-oriented languages, they are still found in purely functional languages and come in handy.

Looking for a match

The first function to look at is the `Any` function. At times, you may be interested in knowing whether or not a certain value is present in a list without being interested in exactly how often it is present or actually using the values afterward. If this is the case, the `Any` function is exactly what you are looking for.

Without the `Any` function, the same result could be achieved somewhat ad hoc with the `Filter` function. You would probably end up writing something such as the following:

```
func main() {
    input := []int{1, 1, 3, 5, 8, 13, 21, 34, 55}
    filtered := Filter(input, func(i int) bool { return i ==
        55 })
    contains55 := len(filtered) > 0
    fmt.Printf("%v\n", contains55)
}
```

Do note that I am splitting this into multiple lines for clarity, but in less verbose languages such as Python and Haskell, this kind of filter would still be a good one-liner. In Go, I'd be a bit cautious about the line length in case you decide to do so.

This implementation has one major flaw. What if you have a really large list of 10 million elements? The `Filter` function will iterate through every element in the list. It is always running in linear time, $O(n)$. Our `Any` function can do better, although we'll still be running in $O(n)$ – worst-case time. In practice, it can be more performant, however.

> **Note**
>
> If we knew that we only needed to look for integers, there are better algorithms than our `Any` implementation here. However, we want to write it generically for any type of data, so those other algorithms would fail for data types such as strings or custom structs.

The easiest way to gain some performance, despite having a theoretical worst-case complexity of linear time, is by iterating through a slice until the first element matches our search. If the match is found, we return `true`. Otherwise, we return `false` at the end of our function:

```
func Any[A any](input []A, pred Predicate[A]) bool {
    for _, element := range input {
        if pred(element) {
            return true
        }
    }
    return false
}
```

Looking for all matches

The implementation for `All` matches is similar to the `Any` match, with the same benefit of abstracting the implementation of `if` statements. The implementation for `All` has a similar practical benefit as the `Any` implementation. As soon as an element does **not** match what we are looking for, we return `false`. Otherwise, we return `true` at the end of our function:

```
func All[A any](input []A, pred Predicate[A]) bool {
    for _, element := range input {
        if !pred(element) {
            return false
        }
    }
    return true
}
```

Implementing DropWhile and TakeWhile

The next two implementations are still predicate-based, but rather than returning a single `true` or `false` as output, these are used to manipulate the slice. In that sense, they are closer to the original `Filter` implementation, but the difference is that they truncate either the start of a list or the tail of a list.

TakeWhile implementation

TakeWhile is a function that will take elements from the input slice as long as a condition is met. As soon as the condition fails, the result containing the start of the list up until the failing predicate is returned:

```
func TakeWhile[A any](input []A, pred Predicate[A]) []A {
    out := []A{}
    for _, element := range input {
        if pred(element) {
            out = append(out, element)
        } else {
            return out
        }
    }
    return out
}
```

In this function, this is exactly what is happening. As long as our predicate is met for each subsequent element, this element is stored in our output value. Once the predicate fails a single time, the output is returned. Let's demonstrate this with a simple slice containing consecutive numbers. Our predicate will look for odd numbers. Hence, as long as the numbers are odd, they will be appended to the output slice, but as soon as we encounter an even number, what we have collected thus far will be returned:

```
func main() {
    ints := []int{1, 1, 2, 3, 5, 8, 13}
    result := TakeWhile(ints, func(i int) bool {
        return i%2 != 0
    })
    fmt.Printf("%v\n", result)
}
```

In this example, the output is [1 1]. Notice how this is different from the plain Filter function – if this same predicate was given to the Filter function, our output would be [1 1 3 5 13].

Implementing DropWhile

Implementing `DropWhile` is the counterpart to `TakeWhile`. This function will drop elements as long as a condition is met. Thus, elements are returned from the first failed predicate test until the end of the list:

```
func DropWhile[A any](input []A, pred Predicate[A]) []A {
    out := []A{}
    drop := true
    for _, element := range input {
        if !pred(element) {
            drop = false
        }
        if !drop {
            out = append(out, element)
        }
    }
    return out
}
```

Let's test this out against the same input data as our `TakeWhile` function:

```
func main() {
    ints := []int{1, 1, 2, 3, 5, 8, 13}
    result := DropWhile(ints, func(i int) bool {
        return i%2 != 0
    })
    fmt.Printf("%v\n", result)
}
```

The output of this function is `[2 3 5 8 13]`. The only elements that are dropped are therefore `[1 1]`. If you combine the output of `TakeWhile` and `DropWhile`, given the same predicate, you would recreate the input slice.

Map/transformation functions

The next category of functions which we will look at is `Map` functions. These are functions that apply a transformation function to each element in a container, changing the element and possibly even the data type. This is one of the most powerful functions in a functional programmer's toolbox, as this allows you to transform your data according to a given rule.

There are two main implementations that we will look at. The first implementation is the simple `Map` function, whereby an operation is performed on each element, but the data type remains the same before and after the transformation – for example, multiplying each element in a slice. This will change the content of the values, but not the type of the values. The other implementation of `Map` is one whereby the data types can change as well. This will be implemented as `FMap`, and this is what we introduced in the previous chapter when looking into Monads.

Transformations while maintaining the data type

The first transformation function that we will look at is one whereby the data types remain the same. Whenever a programmer encounters this function, they can be assured that the data type after calling the function is the same as the data type that was passed to the function. In other words, if the function is called for a list of elements with a data type of `Dog`, the output of this function is still a list of `Dog` elements. What can be different though is the actual content of the fields within those structs (e.g., the name property can be updated).

Just like with the `Filter` implementation, these will be implemented in a purely functional way. Calling the `Map` function should **never** make changes in place to the objects that we provide as an input to the function.

Overall, implementing the `Map` function is straightforward. We will iterate over our slice of values and call a transformation function for each value. Essentially, what we are doing with the `Map` function is abstracting the actual transformation logic. The core algorithm is the iteration over our slice, not the concrete transformations. This means we are once again building a higher-order function:

```
type MapFunc[A any] func(A) A
func Map[A any](input []A, m MapFunc[A]) []A {
    output := make([]A, len(input))
    for i, element := range input {
        output[i] = m(element)
    }
    return output
}
```

In this example, our generic type signature tells us that the data type is preserved when calling `MapFunc`:

```
type MapFunc[A any] func(A) A
```

Given A, we will get A back. Notice that the type can be any type as per the generic contract. There are no type constraints necessary for our Map implementation. Let's look at a demo of multiplying each element in our slice by 2:

```go
func main() {
    ints := []int{1, 1, 2, 3, 5, 8, 13}
    result := Map(ints, func(i int) int {
        return i * 2
    })
    fmt.Printf("%v\n", result)
}
```

This function can work with any data type as well. Let's look at a demo in which we apply a transformation to the name of each dog in a list. If the gender of the dog is male, we'll prefix the name with Mr., and if the gender is female, we'll prefix it with Mrs.:

```go
func dogMapDemo() {
    dogs := []Dog{
        Dog{"Bucky", Havanese, Male},
        Dog{"Tipsy", Poodle, Female},
    }
    result := Map(dogs, func(d Dog) Dog {
        if d.Gender == Male {
            d.Name = "Mr. " + d.Name
        } else {
            d.Name = "Mrs. " + d.Name
        }
        return d
    })
    fmt.Printf("%v\n", result)
}
```

Running this code would result in the following output:

```
[{Mr. Bucky 1 0} {Mrs. Tipsy 3 1}]
```

It's important to stress that these changes are made to copies of the data, and not to the original Dog objects.

Transforming from one to many

An adaptation of the Map function is the Flatmap function. This function will map a **single** item into **multiple** results. Those results will then be collapsed back down into a single list. Collapsing a two-dimensional list down into a one-dimensional list is referred to as flattening the list – hence, Flatmap.

The implementation of the function we will use is not as efficient but works well enough for most purposes. For each element in our slice, we are going to call the transformation function, which will transform our single element into a slice of elements. Rather than storing this intermediately as a slice of slices, we will immediately collapse each slice and store the individual elements consecutively in memory:

```
func FlatMap[A any](input []A, m func(A) []A) []A {
    output := []A{}
    for _, element := range input {
        newElements := m(element)
        output = append(output, newElements...)
    }
    return output
}
```

Let's demonstrate this by implementing an example. For each integer, N, in a slice, we are going to turn this into a slice of all integers from 0 up to N. Finally, we are going to return this result as a consecutive slice:

```
func main() {
    ints := []int{1, 2, 3}
    result := FlatMap(ints, func(n int) []int {
        out := []int{}
        for i := 0; i < n; i++ {
            out = append(out, i)
        }
        return out
    })
    fmt.Printf("%v\n", result)
}
```

The output of running this code is as follows:

```
[0 0 1 0 1 2]
```

This is what we have shown in the image. Every single element is turned into a slice, and the slices are then combined. For each element in our input slice, this is what the intermediate output would look like:

```
0:  [0]
1:  [0 1]
2:  [0 1 2]
```

This intermediate output then gets combined into a single slice. Next, let's take a look at the final category of functions that play a crucial role in functional programming languages.

Data reducing functions

The final group we are going to take a look at is *reducer* functions. These are functions that apply an operation to a container of elements and derive a single value from them. Combined with the functions we have seen earlier in this chapter, we can compose the majority of our applications. At least, as far as data manipulation goes. There are a few different names for functions such as this in functional programming. In Haskell, you'll find functions named `Fold` or `Fold` + a suffix, such as `Foldr`, while in some languages they are called `Reduce`. We will use the `Reduce` terminology for the remainder of this book.

The first function we will look at is simply `Reduce`. This higher-order function abstracts operations to two data elements of the list. It then repeats this operation, accumulating the result, until a single answer is retrieved. Just as with the `Filter` and `Map` functions, these functions are pure, so the actual input data is never changed.

The abstracted function in this algorithm is a function that takes two values of an identical data type and returns a single value of that data. The result is achieved by performing some operation on them that the caller of the function can provide:

```
type (
        reduceFunc[A any] func(a1, a2 A) A
)
```

This function will ultimately be called iteratively for each element in the slice, storing the intermediate results and feeding those back into the function:

> **Note**
>
> This sounds like recursion but it is not recursive in the implementation in this chapter. We will look at a recursive approach in the next chapter.

```
func Reduce[A any](input []A, reducer reduceFunc[A]) A {
    if len(input) == 0 {
        // return default zero
        return *new(A)
    }
    result := input[0]
    for _, element := range input[1:] {
        result = reducer(result, element)
    }
    return result
}
```

In this example, we are also handling our edge cases. If we get an empty slice, we return the default-nil value of whatever type was passed to our function. If there is only one item in the slice, no operation can be performed, and, instead, we just return that value (by not executing the loop and thus instantly returning the result based on input[0]).

These higher-order function abstracts are how you can combine two elements into a single answer. One possible reducer would be sum reducer, which adds two numbers and returns the result. The following anonymous function is an example of this function:

```
func(a1, a2 A) A { return a1 + a2 }
```

This is an anonymous function that we would pass to Reduce to perform a summation of all elements – but there's one problem with this approach as it is written now. The Reduce function is generic and can take **any** type as input, but the + operator is not defined for every data type. To work around this, we can create a Sum function that calls the reducer internally but tightens the type signature to only allow numbers to be provided as input.

Remember that as there are multiple number data types in Go, we want to be able to use the Sum function for all of these. This can be achieved by creating a custom type constraint for our generic functions. We'll also consider a type alias of Number as valid – this can be achieved by adding the ~ prefix to each type:

```
type Number interface {
        ~uint8 | ~uint16 | ~uint32 | ~uint64 | ~uint |
          ~int8 | ~int16 | ~int32 | ~int64 | ~int |
          ~float32 | ~float64
}
```

Next, we can use this type as a type constraint in a generic function such as the Sum function:

```
func Sum[A Number](input []A) A {
    return Reduce(input, func(a1, a2 A) A { return a1 + a2 })
}
```

There we go – now, we can use this function to return a summation of a slice of numbers, whereby a number is any currently supported number-like data type in Go that we have defined in our constraint:

```
func main{
        ints := []int{1, 2, 3, 4}
        result := Sum(ints)
        fmt.Printf("%v\n", result)
}
```

The output of this function is 10. Effectively, our reducer has performed a sum of 1 + 2 + 3 + 4. With the reducer in place, we can therefore abstract these operations to lists. Adding a similar function to perform the multiplication of each element is equally easy to write as the summation function:

```
func Product[A Number](input []A) A {
        return Reduce(input, func(a1, a2 A) A { return a1 * a2
})
}
```

This implementation works the same way as the Sum function.

In Haskell and other functional languages, there are a few different reducer implementations provided out of the box, each changing the core algorithm slightly. You will find the following:

• Reducers that iterate from the start to the end of a list
• Reducers that iterate from the end to the start of a list

- Reducers that start with a default value instead of the first element of a list

- Reducers that start with a default value and then iterate from the end to the start of the list

The reverse reducers (iterating from the end to the start of a list) are left as an exercise for the reader to explore independently, but the full code for them can be found on GitHub: `https://github.com/PacktPublishing/Functional-Programming-in-Go./blob/main/Chapter6/pkg/reducers.go`. However, we will take a look at the reducers that have a starting value.

Providing a different starting value would allow us to write a function such as `multiple all numbers together, and then finally multiply by two`. We could achieve this with some minor modifications to our `Reducer` function:

```
func ReduceWithStart[A any](input []A, startValue A, reducer
reduceFunc[A]) A {
        if len(input) == 0 {
                return startValue
        }
        if len(input) == 1 {
                return reducer(startValue, input[0])
        }
        result := reducer(startValue, input[0])
        for _, element := range input[1:] {
                result = reducer(result, element)
        }
        return result
}
```

We're handling similar edge cases as with the original `Reduce` function, but one key difference is that we always have a default value to return. We can either return it when the slice is empty or return the combination of the starting value with the first element in the slice when the slice contains exactly one element.

In the next example code, we are going to concatenate strings with a comma in between each word, but to show off our new `ReduceWithStart` function, we will provide a starting value of `first`:

```
func main() {
        words := []string{"hello", "world", "universe"}
        result := ReduceWithStart(words, "first", func(s1, s2
          string) string {
                return s1 + ", " + s2
```

```
        })
        fmt.Printf("%v\n", result)
}
```

If we run this code, we will get the following output:

```
first, hello, world, universe
```

With these functions in place, let's take a look at an example in which we combine the use of all three categories of functions.

Example – working with airport data

In this example, we are going to tie together the functions from this chapter to analyze airport data. We need to do some work before we can play around with the functions that we have created. First, we need to get the data. On GitHub, you can find a .json extract under https://github.com/PacktPublishing/Functional-Programming-in-Go./blob/main/Chapter6/resources/airlines.json.

The following snippet is the template for the dataset:

```
{
  "Airport": {
    "Code": string,
    "Name": string
  },
  "Statistics": {
    "Flights": {
      "Cancelled": number,
      "Delayed": number,
      "On Time": number,
      "Total": number
    },
    "Minutes Delayed": {
      "Carrier": number,
      "Late Aircraft": number,
      "Security": number,
      "Total": number,
      "Weather": number
```

```
            }
        }
    }
```

To work with this data, we will recreate the .json structure as structs in Go. We can use the built-in .json tags and deserializers to read this data in memory. Our Go struct to work with this data looks like this:

```
type Entry struct {
    Airport struct {
            Code string `json:"Code"`
            Name string `json:"Name"`
    } `json:"Airport"`
    Statistics struct {
            Flights struct {
                Cancelled int `json:"Cancelled"`
                Delayed   int `json:"Delayed"`
                OnTime    int `json:"On Time"`
                Total     int `json:"Total"`
            } `json:"Flights"`
            MinutesDelayed struct {
                Carrier              int `json:"Carrier"`
                LateAircraft         int `json:"Late
                                     Aircraft"`
                Security             int `json:"Security"`
                Weather              int `json:"Weather"`
            } `json:"Minutes Delayed"`
    } `json:"Statistics"`
}
```

This is a bit verbose, but it's just a copy of what we could find in the first entry of the file. After this, we need to write some code to read this file into memory as entries:

```
func getEntries() []Entry {
        bytes, err := ioutil.ReadFile("./resources/airlines.
            json")
        if err != nil {
                panic(err)
        }
```

```
        var entries []Entry
        err = json.Unmarshal(bytes, &entries)
        if err != nil {
                panic(err)
        }
        return entries
}
```

As in previous chapters, we are using `panic` in this code. It is discouraged, but for demonstration purposes, this is fine. This code will read our resource file, parse it as `json` based on the struct we have created, and return it as a slice.

Now, to demo the functions that we have created, this is what our problem statement looks like: **write a function that returns the total hours of delays for the Seattle airport (airport code: SEA).**

Based on this problem statement, we can see that there are three actions to take:

1.   Filter the data by the airport code SEA.

2.   Transform the `MinutesDelayed` field into hours.

3.   Sum all the hours.

The order of *steps 2 and 3* could be reversed, but this way, it follows the structure in which we have introduced those functions in this chapter:

```
func main() {
    entries := getEntries()
    SEA := Filter(entries, func(e Entry) bool {
            return e.Airport.Code == "SEA"
    })
    WeatherDelayHours := FMap(SEA, func(e Entry) int {
            return e.Statistics.MinutesDelayed.Weather / 60
    })
    totalWeatherDelay := Sum(WeatherDelayHours)
    fmt.Printf("%v\n", totalWeatherDelay)
}
```

And there we go. We have implemented our use case using three of the functions that we have seen in this chapter. As you can tell, whenever we call a function, we store the result in a new slice. The original data is therefore never lost, and we could still use it for other parts of our function should we choose to do so.

Summary

In this chapter, we saw three categories of functions that will help us functionally build our programs. First, we saw predicate-based functions, which can either filter our data into a subset meeting a requirement or tell us whether or not the data in our dataset entirely or partially matches a condition. Next, we saw how data can be changed functionally, ways of transforming data whereby our data type is guaranteed to remain the same, and functions in which we are also changing the type itself.

Finally, we looked at reducer functions, which take a list of elements and reduce them into a single value. We have demonstrated how these three types of functions can be combined in the airport data example.

In the next chapter, we will dive into recursion and see how this plays a role in functional programming, as well as what the performance implications are for writing recursive functions in Go.

7

Recursion

In this chapter, we are going to talk about recursion. This is a topic that all programmers encounter sooner or later, as it's not exclusive to the functional paradigm. Any language in which you can express function calls allows you to express functions that are recursive in nature. For many, it is not a topic that is difficult to understand at first. In functional programming languages such as Haskell, recursion takes center stage.

As such, this chapter is dedicated to understanding exactly how recursion works, including what the performance implications are of doing so, and what the limits of recursion are in Go. We'll also take a look at some handy constructs for dealing with recursion using functions as first-class citizens.

In this chapter, we will cover these main topics:

- What recursion means

- Why use recursive functions?

- When and how to use recursive functions

- Leveraging functions as first-class citizens to write recursive functions

- Understanding the limitations of recursive functions in Go

- Understanding Tail-Recursion and compiler optimizations

What we will learn in this chapter will set us up for success when talking about the Continuation-Passing style and fluent programming in later chapters.

Technical requirements

For this chapter, you should use any version of Go at or above 1.18. All the code can be found on GitHub at https://github.com/PacktPublishing/Functional-Programming-in-Go./tree/main/Chapter7.

What is recursion?

Simply put, a recursive function is a function that calls itself. In practice, this means that the following function is an example of a recursive function:

```
func recursive() {
    recursive()
}
```

In this example, if the user would call the function "recursive," all it would do would call itself ad infinitum. Effectively, this is an infinite loop and not the most useful function. To make recursive functions useful, we can extend our definition of a recursive function a bit further by setting up two rules:

- A function must have a condition on which to call itself (recurse)
- A function must have a condition on which it returns *without* calling itself

The first condition just states that given a function, X, at some point in the function's body, X will be called again. The second condition is that there exists a case for which the function, X, returns from the function without calling itself. This second condition is often called the *base case* of the recursive function.

To understand what this looks like, let's implement a classical mathematical operation that lends itself well to recursion, namely the factorial function. The factorial function is defined as a function that, given an input, *N*, multiplies all the numbers of *N* down to 1; for example:

```
Fact(5) = 5 * 4 * 3 * 2 * 1
```

To see why this is a recursive function, we can show that the result of calling `Fact(5)` is the result of calling 5 multiplied by the result of `Fact(4)`. Thus, if we write this out, we will get the following:

```
Fact(5) = 5 * Fact(4) = 5 * 24 = 120
Fact(4) = 4 * Fact(3) = 4 * 6 = 24
Fact(3) = 3 * Fact(2) = 3 * 2 = 6
Fact(2) = 2 * Fact(1) = 2 * 1 = 2
Fact(1) = 1 * Fact(0) = 1 * 1
Fact(0) = 1
```

Notice that in this example, the factorial of 0 is simply 1. This is defined as our base case; when a value of 0 is passed to our function, we simply return the integer value 1. However, in all other input cases, we are multiplying the input number with the output of calling the factorial function with `input-1`.

If we turn this into Go code, we will get the following:

```go
func main() {
    fmt.Println(Fact(5))
}
func Fact(input int) int {
    if input == 0 {
        return 1
    }
    return input * Fact(input-1)
}
```

If this is the first time you've seen recursion in a while, it may take you a few minutes to wrap your head around what is happening here. One way to think about this is that each function call to Fact pushes a function onto our stack. When all functions are pushed to the stack, they are evaluated from top to bottom, and each lower level of the stack can use the result from what came above it:

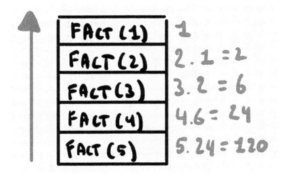

Figure 7.1: Recursive function calls and stack allocation

Thinking about recursion in this stack-based way will help us understand the examples and pitfalls of recursion later in this chapter. But before we get to that, let's look at when you might want to opt for writing a recursive function instead of an iterative one and why functional languages typically favor recursion.

Why do functional languages favor recursion?

Before we discuss when to use recursive functions in Go, let's answer the question of why functional languages seem to prefer recursion rather than `for` loops. The best answer for this is that recursion is inherently purer than iterative solutions. Although each program that can be expressed recursively can also be expressed iteratively, iterative solutions need to maintain more state than recursive solutions.

Our simple factorial example highlights this when we write an iterative implementation:

```
func factorial(n int) int {
    result := 1
    for i := 1; i <= n; i++ {
        result = result * i
    }
    return result
}
```

In this factorial implementation, we are mutating the "result" in each iteration of the `for` loop. It is a well-contained mutation as it does not escape the function itself, but it's a mutating state, nonetheless. Meanwhile, our pure recursive example never mutates the state. Rather than mutating the state, it returns a **new** value by combining an input parameter with the output of a function call:

```
return input * Fact(input-1)
```

As a general rule, recursion allows us to create new functions with copied states, make changes to the copies, and return the result, all without mutating a value in the recursive call itself. This means that changes to the program state are contained within each stack frame.

> **Recursive state changes in Go**
>
> In Go and other non-pure languages, it is possible to mutate the state in recursive function calls. In such languages, recursion does not guarantee the immutability of the state but it does make it easier to write immutable implementations.

When to use recursive functions

To understand when to use a recursive function, we have to talk about the main trade-offs between iterative and recursive functions. But before we get to that, let's start by saying that anything that can be implemented iteratively can also be implemented recursively. As such, each function that has a `for` statement in Go can be replaced by an equivalent function that uses a recursive function call in place of the `for` loop.

However, we might not always want to do so. The two main disadvantages of recursive functions are that they typically have greater time and space requirements. Calling a function multiple times creates multiple stack frames. These stack frames use up part of the working memory of our programs. Typically, each frame will contain a copy of the data from the frame below it (in a recursive function), which means that in the earlier Factorial example, each function call uses a similar amount of memory as the function that came before it. However, all these stack frames, at some point, are alive at the same time. A recursive call stack does not pop the stack until the final recursive call is completed. Hence, in *Figure 7.1*, we can see that all stacks are added on top of each other, and are then evaluated from top to bottom (**Last-In, First-Out**, or **LIFO**). Had we written the same function iteratively, we would only have had one function on the call stack.

The second limitation of recursive functions is that they are typically slower than their iterative counterparts. This is mostly because function calls are expensive operations as far as programming language features go. In light of what we have just learned about the call stack, this makes sense. Each function call has to copy over memory to a new location, perform the core algorithm, and then copy it all over again for the next recursive call.

So, why would we want to still use recursive functions? Well, although these limitations are important, our main goal is to achieve code readability and maintainability. Recursion, once mastered, can make programs not only easier to write but also easier to understand. Problems that involve traversing over graphs or trees easily lend themselves to recursive functions (as these data structures are recursive data structures themselves). An overarching theme of this book is that we'll trade off performance for the convenience of both you, the programmer, and later readers of the code.

As a side note, in languages such as Haskell, writing recursive functions involves less syntax overhead than in Go – especially when combined with a concept known as *pattern matching*. Without diverging too much from the core content of this chapter, let's quickly look at the factorial implementation in Haskell:

```
factorial :: Integral -> Integral
factorial 0 = 1
factorial n = n * factorial (n-1)
```

The preceding snippet is a full implementation of the Factorial function. Notice that it reads almost like a more mathematical description of the problem. This makes writing the recursive solution more appealing. In addition, Haskell also performs compiler-level optimizations for handling recursive functions. We'll briefly look at one such optimization, Tail-Call optimization, later in this chapter.

Iterating over trees

To demonstrate the preceding assumption that some code is easier to write recursively rather than functionally, let's take a look at an example of iterating over a tree. Trees are recursive data structures, and as such should lend themselves to this implementation. For simplicity, let's assume we have a tree that stores integers; the actual values don't matter as much. We'll construct a tree that looks like this:

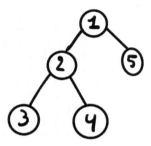

Figure 7.2: Example of a (binary) tree

The actual values of each node don't matter, but let's say we want to find the sum of all nodes. In plain English, what we have to do is get the value of each node. Then, for each node, we need to figure out if it has children. If so, we add the value of the child to our running sum. Next, for all of those children, we figure out if they have children, and if so, also add their values to our running sum. We do this until we have seen all the nodes.

To demonstrate this, let's create a data structure that represents our tree. The type declaration itself is straightforward: we have a node that contains a value, and each node has a pointer to a left and right child. These children are optionally present:

```
type node struct {
        value int
        left  *node
        right *node
}
```

With this struct set up, let's also introduce an actual tree on which we can demonstrate our example functions later in this chapter. We can create this as a package-level object in a var block. We'll model the tree shown in *Figure 7.2*:

```
var (
        ExampleTree = &node{
                value: 1,
                left: &node{
```

```
            value: 2,
            left: &node{
                    value: 3,
            },
            right: &node{
                    value: 4,
            },
        },
        right: &node{
            value: 5,
        },
    }
)
```

Before we write this as a recursive solution, let's write this as an iterative solution using a normal `for` loop.

Iteratively solving tree problems using a for loop

We need to introduce some additional data structures before we can make this work. The data structure that we will use here is a `Queue`. For each node that we visit, we will add the node's value to our sum. For each child of the node, we will add the child to our `Queue`. We will keep doing this until our `Queue` is empty. As a starting value, we will add the root of our tree to our `Queue` to kickstart the entire process.

One important disclaimer is that, at the time of writing, Go does not ship with an easy-to-use, out-of-the-box queue implementation. However, Go does include buffered channels out of the box. We can use buffered channels to get queue-like behavior, which is what we will be doing to demonstrate this. The main properties to get queue-like behavior are as follows:

- Being able to push an element to the queue
- Being able to pop (remove) an element from the queue in LIFO style

You could use a slice to get this behavior, but even that requires some overhead for managing the slice and it's not the most performant implementation. A real queue would offer constant-time addition and removal. For that matter, perhaps buffered channels are doing this in an optimized way under the hood, but further exploration of that is outside the scope of this book. One necessary assumption we have to make, however, is that we know the size of our queue beforehand.

In a real-world scenario, this is often not the case. You could pass a best-effort estimation for the queue size to the buffered channel, but this seems error-prone. For didactic purposes and not to distract from the essence of the algorithm, we will accept those assumptions for now. With this disclaimer out of the way, let's learn how to implement a function to get the sum of all the nodes in a tree iteratively:

```go
func sumIterative(root *node) int {
        queue := make(chan *node, 10)
        queue <- root
        var sum int
        for {
                select {
                case node := <-queue:
                        sum += node.value
                        if node.left != nil {
                                queue <- node.left
                        }
                        if node.right != nil {
                                queue <- node.right
                        }
                default:
                        return sum
                }
        }
}
```

In this example, we are adding a bit of additional overhead since we are managing our queue behavior using buffered channels. However, the core algorithm is the same. You could imagine saving some lines of code by not having a select block when using a real queue implementation though.

Next up, let's take a look at how we can solve this problem recursively.

Recursively solving tree problems

When thinking about this problem recursively, it becomes much clearer and easier to implement.

Remember from our factorial example that we are adding calls to our stack frame until we encounter a base case for which we can return a value without calling the function itself. The base case for this implementation is an absent node (nil pointer). Such a node will return a value of 0 as there is no sum to be made. For each other node, we return the sum of its value, along with the sum of values for all children. Visualizing this like a stack, we are adding frames to our stack from bottom to top, but evaluating from top to bottom, aggregating the sum as we go along:

```
func sumRecursive(node *node) int {
        if node == nil {
                return 0
        }
        return node.value + sumRecursive(node.left) +
            sumRecursive(node.right)
}
```

This recursive code is one way to solve this problem without too much overhead. It is a more readable version of the iterative solution, and our code is closer to our intention. How does the recursive solution relate to what we have learned about functional programming so far?

In functional programming languages, you want to tell the computer *what* to solve instead of *how* to solve it. When you are writing loops manually, you are firmly in the domain of the how rather than the what of a given problem. In addition, our recursive solution is not mutating state anywhere, which brings us closer to an ideal function in the world of functional programming.

> **Functional languages and loops**
>
> While recursion is preferred in functional languages, many do offer constructs for creating manual loops as well. That said, they often offer compiler optimizations for recursive functions, which makes them an even more attractive option to solve problems.

Recursion and functions as first-class citizens

What we have seen so far in this chapter can be applied to any language that has function calls, even in languages that stick more firmly to the object-oriented domain. In this section, we'll learn how to leverage some of the concepts of functional and multi-paradigm languages that make recursion easier to write and manage.

One of the most useful features I've found is to combine recursion with closures. To give an example of when this comes in handy, imagine working recursively on a data structure and having to keep some state tracked. Rather than tracking the state at the package level, or complicating the recursive function to keep the state tracked in the recursing functions, *we can create an outer function that is not recursive and then use a recursive inner function*. Let's demonstrate this with an example to clear up some potential confusion.

Using the same tree as in the previous example, let's write a function to find the maximum value of a node in the tree. To achieve this, we need a way to track what the maximum value is, which we've seen so far. One option to achieve this is by tracking the state in a global variable outside of our recursive function. This is messy but would work. For example, the following code traverses the tree and uses a global variable to track what the maximum encountered value is as follows:

```
var maximum = 0
func MaxGlobalVariable(node *node) {
        if node == nil {
                return
        }
        if node.value > maximum {
                maximum = node.value
        }
        MaxGlobalVariable(node.left)
        MaxGlobalVariable(node.right)
}
func main() {
        maximum = int(math.MinInt)
        MaxGlobalVariable(ExampleTree)
        fmt.Println(maximum)
}
```

The preceding code is not the ideal solution. First of all, using global variables to track any state should be discouraged. It would cause major headaches when writing multithreaded code, and if you'd forget to reset the global variable before a run of the recursive function. The outcome would be unreliable, even for single-threaded runs.

Another much better approach is to track the current maximum value as part of each recursive call. This is achieved by extending the function signature so that it includes the integer value that we are tracking, as shown in the following code:

```
func.maxInline(node *node,
    maxValue int) int {
        if node == nil {
                return maxValue
        }
        if node.value > maxValue {
                maxValue = node.value
```

```
        }
        maxLeft := maxInline(node.left, maxValue)
        maxRight := maxInline(node.right, maxValue)
        if maxLeft > maxRight {
                return maxLeft
        }
        return maxRight
}
```

Here, we are tracking the maximum value in the maxValue variable, which is passed in each recursive call. Then, in each call, we are continuing the recursive call downwards with the maximum value between node.value and maxValue. We end the calls by comparing the left and right-hand sides of the tree and returning the max of both sides.

This is probably the cleanest way of writing the recursive function itself if we ignore what the code of the caller looks like. If we want to call the maxInline function, our calling functions will look like this:

```
func main() {
        fmt.Println(maxInline(ExampleTree, 0))
}
```

In the function call to maxInline, we are effectively leaking an implementation detail to the caller. The caller has to pass the initial starting value to our recursive function. This is rather messy, and for more complex functions, we can't necessarily expect the caller to know what the appropriate value is. Ideally, we don't leak such state details to our callers. Traditional object-oriented languages solve this problem by exposing a public non-recursive function that calls a private recursive function with the state appended. Modeling this in Go, we get the following:

```
func main() {
        fmt.Println(MaxInline(ExampleTree))
}
func MaxInline(root *node) int {
        return maxInline(root, 0)
}
func maxInline(node *node, maxValue int) int {
        if node == nil {
                return maxValue
        }
        if node.value > maxValue {
                maxValue = node.value
```

```
        }
        maxLeft := maxInline(node.left, maxValue)
        maxRight := maxInline(node.right, maxValue)
        if maxLeft > maxRight {
                return maxLeft
        }
        return maxRight
}
```

Here, we have created a public `MaxInline` function that does not expose the internal mechanism for `maxInline`. The caller only needs to provide the root node to the public function. This function will then call the private `maxInline` function with the appropriate starting state. This pattern is incredibly common in object-oriented languages, and if those languages don't support first-class functions, this is the right way to go about it.

However, in Go, we can do better. The main issue with the preceding approach is that you are still cluttering the package-private space with a function anyone working in the package can use. This might be desired behavior, but not always. One way to work around this is by encapsulating the recursive function *within* the non-recursive function. In this way, we can track the state inside the non-recursive function, which is accessible to the recursive inner function.

The following implementation does exactly that:

```
func Max(root *node) int {
        currentMax := math.MinInt
        var inner func(node *node)
        inner = func(node *node) {
                if node == nil {
                        return
                }
                if node.value > currentMax {
                        currentMax = node.value
                }
                inner(node.left)
                inner(node.right)
        }
        inner(root)
        return currentMax
}
```

Let's take a look at what is happening here. First, note that our Max function is not recursive itself. This allows us to perform some operations that we know will only happen once per call to Max. For example, this is a great place to log activity, add metrics for performance, or add some state, as we are doing here. In our case, we're creating a variable called currentMax. This variable will keep track of what the maximum value is that we've encountered.

Next, we are creating a variable called inner of the func(node *node) type. This is an important step. We're not creating the function in-line immediately; first, we need to set up the variable without an implementation attached to it. The reason why we are doing this is so that we can refer to the inner variable inside an anonymous function.

The next step is to instantiate this inner function. If we tie that block together, we get this:

```
var inner func(node *node)
inner = func(node *node) {
        if node == nil {
                return
        }
        if node.value > currentMax {
                currentMax = node.value
        }
        inner(node.left)
        inner(node.right)
}
```

This shows how we are calling inner(node.left) and inner(node.right) from within the inner function itself. This would not work if we did not define the function first without instantiating. In other words, the following code would not work:

```
inner := func(node *node) {
    if node == nil {
        return
    }
    if node.value > currentMax {
        currentMax = node.value
    }
    inner(node.left)
    inner(node.right)
}
```

It's a seemingly small change, but it would break our function. After all, how could we refer to `inner` if the compiler hadn't yet compiled the function that you are trying to create?

The last step of our code is to call the inner recursive function itself:

```
inner(root)
```

In this example, we have seen how using functions as first-class citizens can help us write recursive code. But there are performance implications of doing so. We'll explore this in the next section.

Limits of recursive functions

Recursive functions have a performance penalty. When creating a recursive function call, we are copying over the state from one function stack to the next. This involves copying a lot of data into our working memory, but additional computational overhead is required to make the function call itself happen. The main limitation of solving problems recursively, at least in Go, is that we will eventually run out of space to make the recursive call happen. The other limitation is that a recursive solution is often slower than an iterative one.

Measuring the performance of recursive versus iterative solutions

Before we look at the implications for the space our programs are using during recursive function calls, let's compare the performance of recursive and iterative solutions that fit within our working memory. To demonstrate this, we will use the same iterative and recursive solution to the factorial problem that we saw at the start of this chapter:

```go
package pkg
func IterativeFact(n int) int {
        result := 1
        for i := 2; i <= n; i++ {
                result *= i
        }
        return result
}
func RecursiveFact(n int) int {
        if n == 0 {
                return 1
        }
        return n * RecursiveFact(n-1)
}
```

To test both functions, we can use the benchmarking features of Go, which we explored in earlier chapters. The benchmark setup for both the iterative and recursive approach is straightforward:

```
package pkg
import "testing"
func BenchmarkIterative100(b *testing.B) {
        for n := 0; n < b.N; n++ {
                IterativeFact(10)
        }
}
func BenchmarkRecursive100(b *testing.B) {
        for n := 0; n < b.N; n++ {
                RecursiveFact(10)
        }
}
```

To benchmark the functions, we are going to generate the result of `Factorial(10)`. This is a pretty low number as it takes only 10 steps to derive the answer. Yet, the performance implications are clear. The average of multiple runs is as follows:

Function	ns/op
Iterative	8.2
Recursive	24.8

Table 7.1: Performance of iterative versus recursive functions in ns/op

As we can see, each iterative function needed about 1/4th the time to complete compared to the recursive function. The following graph shows the runtime of each function in ns/op for different inputs to the factorial function:

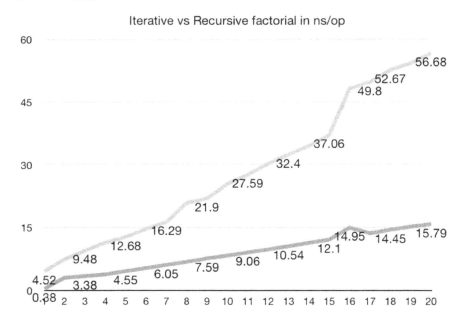

Figure 7.3: Iterative (bottom) versus recursive (top) runtime in ns/op

The preceding graph shows that not only are recursive functions typically slower than their iterative counterparts, but they become slower in a more drastic way than the iterative solution. Keep these performance considerations in mind when opting to write recursive functions.

Note on benchmarking

These results were obtained using an Amazon Web Services EC2 instance (t2.micro) running Amazon Linux. The actual values of these results are machine-dependent. Running these benchmarks on a different machine will not necessarily give different results, but the general trend should remain the same. Running the benchmarks on the same t2.micro instance can still cause variations in the outcome.

Space limitation of recursive functions

Apart from being slower in a typical scenario, recursive functions suffer from another drawback: each function called to a recursive function adds another frame to our stack. All the data from the current iteration is copied over and passed on to the new function. Recall from *Figure 7.1* that these stacks are added on top of each other in a **LIFO** fashion. Once our stack cannot grow any further, the program will halt. The good news is that in Go, this limit is relatively large and might not pose immediate practical problems. On a modern 64-bit machine, this stack can hold up to 1 GB of data, while on 32-bit machines, the limit is 250 MB.

In practice, the limits will eventually get hit. Let's take a look at the following example:

```
func main() {
        infiniteCount(0)
}
func infiniteCount(i int) {
        if i%1000 == 0 {
                fmt.Println(i)
        }
        infiniteCount(i + 1)
}
```

If we were to run this function on a 32-bit machine, the tail end of the output would look like this:

```
1861000
1862000
1863000
1864000
runtime: goroutine stack exceeds 262144000-byte limit
runtime: sp=0xc008080380 stack=[0xc008080000, 0xc010080000]
fatal error: stack overflow
runtime stack:
runtime.throw({0x496535?, 0x50e900?})
        /usr/lib/golang/src/runtime/panic.go:992 +0x71
runtime.newstack()
        /usr/lib/golang/src/runtime/stack.go:1101 +0x5cc
runtime.morestack()
        /usr/lib/golang/src/runtime/asm_amd64.s:547 +0x8b
```

Hence, after about 1.8 million iterations, our program will crash. The actual limit depends on how large each stack frame is. For recursive functions that are more complex and manage more internal state, this limit will be lower. But what can we do to avoid hitting this limit? In Go, there is no way to completely avoid this limit when dealing with recursive functions. However, we can adjust the limit (although the 1 GB limit on a 64-bit machine should be plenty).

To alter the limit, we can use the debug.SetMaxStack(bytes) function. To demonstrate this, let's alter the limits of a 32-bit machine to be twice the default size:

```
func main() {
        debug.SetMaxStack(262144000 * 2)
```

```
        infiniteCount(0)
}
func infiniteCount(i int) {
        if i%1000 == 0 {
                fmt.Println(i)
        }
        infiniteCount(i + 1)
}
```

Now, the function can go on much longer before running out of stack space:

```
3724000
3725000
3726000
3727000
3728000
runtime: goroutine stack exceeds 524288000-byte limit
runtime: sp=0xc010080388 stack=[0xc010080000, 0xc020080000]
fatal error: stack overflow
runtime stack:
runtime.throw({0x496535?, 0x50e900?})
        /usr/lib/golang/src/runtime/panic.go:992 +0x71
runtime.newstack()
        /usr/lib/golang/src/runtime/stack.go:1101 +0x5cc
runtime.morestack()
        /usr/lib/golang/src/runtime/asm_amd64.s:547 +0x8b
```

As we can tell, we could complete about 3.7 million iterations now before running into the limits of a 500 MB stack. While the 250 MB limit on a 32-bit machine is not extensive, for most practical applications, the 1-GB limit on a 64-bit machine should be sufficient.

Tail recursion as a solution to stack limitations

Considering these limitations of recursive functions, it might seem strange that functional languages prefer recursion over iteration. Often, these languages, such as Haskell, only have recursion to work with, and they mock iterative functions. In this section, we will briefly look at how languages such as Haskell make recursion work.

> **Tip**
> The important thing to note here is that this is not possible in Go at the time of writing.

The technique some functional languages use is called **tail-call optimization**. Even non-functional languages might offer this – JavaScript is a notable example. This is a compiler (or interpreter) optimization whereby a recursive function call is made without allocating a new stack frame. Recall that the main drawback of recursive functions is the fact that they can run out of stack space – hence, if we solve that problem, we can have infinite recursion.

The compiler does need some help from the programmer to make this work. We'll demonstrate the examples with Go, but keep in mind that so far in Go, the compiler performs no optimization and as such we would still overflow the stack eventually.

Rewriting a recursive function into a tail-call recursive function

The key difference between a tail-call recursive function and a normal recursive function is that in the tail-call variant, each stack frame is independent of the others. To show this, let's examine the factorial function again:

```
func Fact(input int) int {
    if input == 0 {
        return 1
    }
    return input * Fact(input-1)
}
```

In the last line of this function, we are returning input * Fact(input - 1). This effectively ties the result of each call to the result of the subsequent call. To evaluate the multiplication, we'd first have to run the Fact function one level deeper. We could rewrite this function to avoid this and make each stack frame independent of the next.

To do this, let's leverage our functions as first-class citizens again. We'll create an outer function called tailCallFactorial that is non-recursive, which, in turn, calls an inner function called factorial, which is recursive.

To write this function recursively and decouple each stack frame, we'll make two changes. First, we'll use a counter that counts down from input to 0. This is equivalent to the for i := n; i > 0; i- for loop. Next, we'll also keep aggregating the result of each multiplication. We will do this by performing multiplication on the input arguments of the next frame and passing on the multiplied values:

```
func tailCallFactorial(n int) int {
    var factorial func(counter, result int) int
    factorial = func(counter, result int) int {
```

```
        if counter == 0 {
            return result
        }
        return factorial(counter-1, result*counter)
    }
    return factorial(n, 1)
}
```

The key line of code that makes this function tail-call recursive is as follows:

```
return factorial(counter-1, result*counter)
```

With this simple change, each stack frame can be evaluated separately. And some compilers detect that the current stack frame can be de-allocated as soon as the next frame is called. This is a high-level overview of what tail-call optimization is, but keep in mind that Go does not perform such compiler optimizations at the time of writing.

Summary

In this chapter, we saw why recursion is a critical part of functional programming languages. We looked into how recursive functions make it easier to enforce function purity and immutability. Next, we saw how functions as first-class citizens can make it easier to manage the state of our recursive function calls. We did this by creating outer non-recursive functions that leverage an inner recursive function to perform the calculations.

After, we looked into the performance concerns of recursive and iterative solutions. Here, we saw that recursive solutions are often slower than their iterative counterparts and that eventually, recursive functions run out of memory to operate with, causing our programs to halt (even though this would take a very long time on a 64-bit machine).

Finally, we looked at Tail-Call optimization and Tail-Call recursive functions. Tail-Call optimization is a practical compiler optimization that many languages, such as Haskell and JavaScript, support to work around the limitations of recursive functions. Crucially, we have seen that Go does not support Tail-Call optimization, even if we write Tail-Call recursive functions.

In the next chapter, we will look at declarative and fluent programming. We'll leverage recursion to write programs in a continuation-passing style.

Readable Function Composition with Fluent Programming

In this chapter, we are going to look at different methods for chaining functions in functional programming. The end goal here is to write code that is easier to read and takes up less visual space. We will look at three ways to achieve this:

- First, we will take a look at how we can use type aliases to attach methods to container types, allowing us to create chained functions with the familiar *dot notation*.

- We'll then discuss lazy versus eager code evaluation.

- Next, we will take a look at **continuation-passing style (CPS)** programming. In CPS, we will use higher-order functions to create a control flow without our functions.

- We'll also discuss the trade-offs in each approach.

Technical requirements

For this chapter, the minimum requirement is Go 1.18, as we will be writing code using generics. All the code can be found on GitHub: `https://github.com/PacktPublishing/Functional-Programming-in-Go./tree/main/Chapter8`.

Some of the code in this chapter will build on top of functions created in both *Chapter 5* and *Chapter 6*. Where necessary, I have copied over the relevant functions and types from those chapters into the `Chapter8` subfolder. For example, `Chapter8/LazyEvaluation/pkg` is a copy of `Chapter5/Monads/pkg` and `Chapter6/pkg`. This way, the examples in `Chapter8` can always be run without requiring the other chapters.

Chaining functions through dot notation

Chaining functions through dot notation is not a unique concept to functional programming. In fact, many object-oriented patterns such as the builder pattern explicitly do this as well. Before we dive into how we can leverage Go's type aliases to do this, let's look at an example in a more object-oriented style of programming before we dive into chaining functions.

Chaining methods for object creation (builder pattern)

We will create a package-private `person` object and add some public functions to change the state of the person, although remember that in Go, this is not the best way of instantiating a new object. However, it is the method many traditional object-oriented languages opt for:

```
type person struct {
        firstName string
        lastName  string
        age       int
}
func newPerson() *person {
        return &person{}
}
func (p *person) SetFirstName(firstName string) {
        p.firstName = firstName
}
func (p *person) SetLastName(lastName string) {
        p.lastName = lastName
}
func (p *person) SetAge(age int) {
        p.age = age
}
```

In this example, we have a person struct and three setters – `SetFirstName`, `SetLastName`, and `SetAge`. All three are used to modify the state of our object. If we want to create a new object, we can do so with the following function calls:

```
func main() {
        alice := newPerson()
        alice.SetFirstName("alice")
        alice.SetLastName("elvi")
```

```
        alice.SetAge(30)
        fmt.Println(alice)
}
```

Alternatively, a constructor can be created:

```
func constructor(firstName, lastName string, age int)
    person {
            return person{firstName, lastName, age}
}
```

This approach would work fine as long as our objects contain few fields. If an object contains many fields, the constructor and setter approach becomes error-prone and, frankly, tedious to write and maintain. When some fields need default values, it becomes even harder to model in many traditional languages (although some, such as Python and TypeScript, deal with this scenario gracefully). A solution to this particular problem is the **builder pattern**, which allows you to chain function calls for a more readable object-creation experience. It also offers additional benefits, such as being able to define default values, but for the purpose of this chapter, we'll just focus on chaining method calls.

To achieve this, we will create a new type, `personBuilder`, which has a function for each field that we want to set. However, rather than simply mutating the `person` object, we will return `personBuilder` with the changes applied. Recall from earlier chapters that this is a way to ensure our functions are pure. It also allows us to create these functions without having to use pointers, as our state will be immutable. The one additional function that we need is `build()`, which will return the fully instantiated object:

```
type personBuilder struct {
        person
}
func (pb personBuilder) FirstName(firstName string)
  personBuilder {
        pb.person.firstName = firstName
        return pb
}
func (pb personBuilder) LastName(lastName string)
  personBuilder {
        pb.person.lastName = lastName
        return pb
}
func (pb personBuilder) Age(age int) personBuilder {
```

```
        pb.person.age = age
        return pb
}
func (pb personBuilder) Build() person {
        return pb.person
}
```

When we want to create a person using `personBuilder`, we can chain the functions using the familiar dot notation:

```
func main() {
        bob := personBuilder{}.FirstName("bob").
          LastName("Vande").
          Age(88).
          Build()
        fmt.Println(bob)
}
```

Dot notation to chain functions on slices

With this brief reminder of how dot notation works and how it is used in object-oriented languages, let's dive into how we can leverage the same concept for the functions that are encountered in functional programming languages. Recall from earlier chapters that we created functions such as `filter`, `map`, and `sum` (as an abstraction on top of `reduce`). When we wanted to run multiple functions in sequence, we had to do so in separate statements and keep track of values in between. For example, imagine we have a slice of numbers. We want to double each number, then keep only those larger than 10, and finally, return their sum. Using the functions of *Chapter 6*, we could write this as follows:

```
func main() {
        ints := []int{1, 2, 3, 4, 5, 6, 7, 8, 9, 10}
        doubled := Map(ints, func(i int) int { return i * 2 })
        larger10 := Filter(doubled, func(i int) bool {
          return i >= 10 })
        sum := Sum(larger10)
        fmt.Println(sum)
}
```

Technically, we don't need the intermediate steps. We can write it as a one-liner, but it becomes incomprehensible rather quickly:

```go
func oneliner() {
        ints := []int{1, 2, 3, 4, 5, 6, 7, 8, 9, 10}
        sum := Sum(Filter(Map(ints, func(i int) int {
            return i * 2 }), func(i int) bool {
            return i >= 10 }))
        fmt.Println(sum)
}
```

With some minor formatting changes, it becomes slightly more readable, but it's still not great, although it has a bit of a Lisp-y feel to it:

```go
func oneliner() {
        ints := []int{1, 2, 3, 4, 5, 6, 7, 8, 9, 10}
        sum := .Sum(
                .Filter(
                    .Map(ints,
                        func(i int) int {
                            return i * 2 }),
                func(i int) bool { return i >= 10 }))
        fmt.Println(sum)

}
```

If you spend some time reading functions such as the preceding example, you do get used to it. Common Lisp is a good example here; the parenthesis makes it hard to read initially, but over time, it becomes second nature. Yet, I'd argue most of your coworkers are not fluent Lisp programmers and likely don't want to spend their time learning how to read such code. As object-oriented dot notation is the most common way of method chaining, we should opt for a solution that is closer to what the majority of people are used to. We can achieve this in Go using type aliases. Remember from *Chapter 2* that type aliases allow us to attach functions to custom types and that we can create a custom type to represent a slice.

The first step, then, is to create a type alias for our container type. This works for all types, but we'll demonstrate it with integers:

```go
type ints []int
```

Next, we will attach custom methods to this type alias. For our example, we will use `Map`, `Filter`, and `Sum`, as in the previous example, but this would work for any function. For each of the functions, they will call our existing (generic) `Map`, `Filter`, and `Sum` methods. However, notably, the difference is that these functions are now attached to a concrete type. This is somewhat similar to creating a *façade* pattern for function dispatching:

```go
func (i ints) Map(f func(i int) int) ints {
        return .Map(i, f)
}
func (i ints) Filter(f func(i int) bool) ints {
        return Filter(i, f)
}
func (i ints) Sum() int {
        return .Sum(i)
}
```

As you can tell from the preceding snippet, there's no real magic happening here, but this small change will allow us to chain our functions together in the familiar dot notation. For example, the following method is identical to the previous non-chained examples:

```go
func chaining() int {
        input := ints([]int{1, 2, 3, 4, 5, 6, 7, 8, 9, 10})
        return input.Map(func(i int) int { return i * 2 }).
                Filter(func(i int) bool { return i >= 10 }).
                    Sum()
}
```

I'd wager that, for many people, this is the more readable version, especially compared to the more Lisp-style example. However, to an extent, this is just personal preference and what you are used to. That said, in the population of Go programmers, dot notation function chaining is the more common approach. The main downside of this approach is that new functions need to be created simply to allow dot notation chaining. The good news is that there are solutions available for this, but they will make your project setup a bit more complex. We can automate the generation of such functions for our types using the Go compiler pragma system. In *Chapter 11*, we will see some examples of libraries that can do this.

Lazy evaluation of function calls

There is a trade-off that happens whenever we opt for the preceding dot notation style of declarative programming in Go. To understand why there is a potential negative performance impact when chaining functions in Go but not in a language such as Haskell, we need to understand the concept of function evaluation and, particularly, lazy evaluation. When a programming language is said to support **lazy evaluation** of a function call, what is meant is that the function is only executed at the moment when the result is needed, instead of ahead of time.

We can contrast this with **eager evaluation** (also called strict evaluation), where the entire result for each function is computed at the time of the function call. Eager evaluation is the execution strategy employed by most programming languages, so it is likely the one that you are most familiar with. *Go does not opt for lazy evaluation, but we can mimic it.* To understand what it means for a programming language to be lazy, let's first talk about eager evaluation and the mental model associated with this. Think about the flow of execution for the following snippet:

```
func main() {
        x := 3
        y := 4
        z := x + y
        fmt.Println(z)
}
```

When reading this code, the flow of execution follows our way of reading pretty much. First, the top line is evaluated, and then the one below that, all the way to the final line in the code.

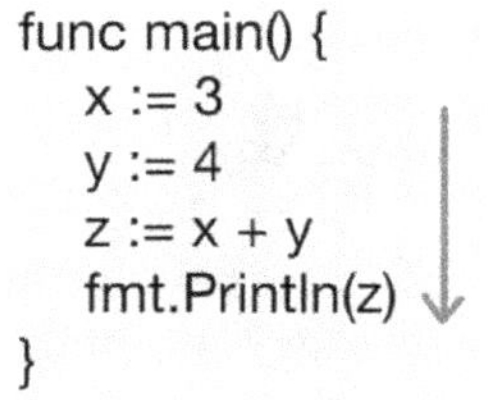

Figure 8.1: The flow of execution, from top to bottom

This is a natural way to read code and follow what is happening. Let's extend the example with some function calls.

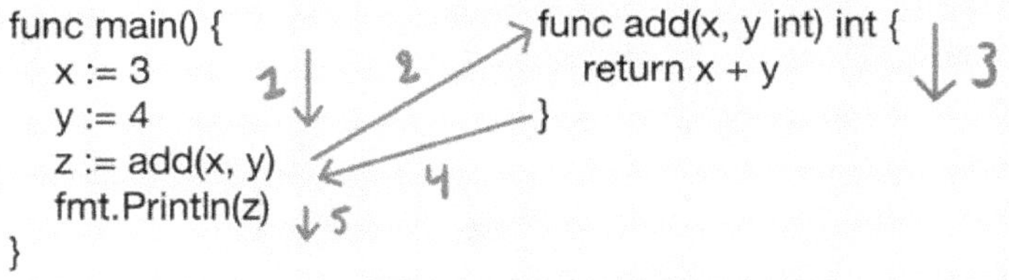

Figure 8.2: The execution flow with function calls

In *Figure 8.2*, we can see how the execution flow is modeled when a function call is present. First, **x** and **y** are initialized. Then, before the assignment to **z** can happen, the **add** function needs to be executed entirely (line three in the image). Finally, the flow of execution is handed back over to our `main` function, which ends up printing the result stored in **z**. This is pretty straightforward and likely how you have been thinking about execution flow all along. Let's show one more example of eager evaluation, and then contrast it with lazy evaluation. For this example, we will use the `Filter` method that we created in *Chapter 6*:

```
func main() {
    input := []int{1, 2, 3, 4, 5, 6}
    isEven := func(i int) bool {
        return i%2 == 0
    }
    numberPrinter(pkg.Filter(input, isEven))
}
func numberPrinter(input []int) {
    for _, in := range input {
        fmt.Println(in)
    }
}
```

With eager evaluation, what happens in the preceding snippet is that the call to `Filter` will be resolved before passing on the entire result to `numberPrinter`. Essentially, the most deeply nested function will be evaluated first, with the outermost function evaluated last (and using the result of the inner evaluation). Again, this is how most of us rationalize code. *Lazy evaluation, however, wants to only perform the computation once the result becomes required.* In the preceding example, the moment at which the "even number filter" becomes relevant is when we start iterating over the result in `numberPrinter`. Thus, the flow of execution looks like *Figure 8.3*.

```
func main() {                                    func numberPrinter(input []int) {
    input := []int{1, 2, 3, 4, 5, 6}                 for _. in := range input {
    isEven := func(i int) bool {                         fmt.Println(in)
        return i%2 == 0                              }
    }                                            }
    numberPrinter(pkg.Filter(input, isEven))
}                                                func Filter[A any](input []A, pred Predicate[A]) []A {
                                                     output : = []A {}
                                                     for _, element := range input {
                                                         if pred(element) {
                                                             output = append(output, element)
                                                         }
                                                     }

                                                     return output
                                                 }
```

Figure 8.3: The lazy evaluation execution flow

In *Figure 8.3*, we zoom in on what happens once the numberPrinter(pkg.Filter(input, isEven)) line is reached. What happens during lazy evaluation is that we jump into the numberPrinter function. Because the filtered list of numbers is not yet relevant to enter that function, the call to pkg.Filter has not yet happened. However, our runtime makes note that this function will need to be called eventually. Next, we hit the first line of numberPrinter, which loops over our input. At this point, the result of the Filter function becomes relevant. Thus, we need to figure out which numbers are odd by calling pkg.Filter. Once the result has been computed, the execution continues at the [..] range input [..] line. Thus, execution was effectively deferred until it was needed. That is the critical point of lazy evaluation – no work (i.e., no processing power) is expended until we know that it will be absolutely necessary.

A language built around this strongly requires function purity, as having a shifting state of the system in conjunction with this lazy evaluation mode of execution would be a recipe for disaster, and would be a leading cause of headaches among functional programmers. Go does not automatically translate our code into functions called with lazy evaluation, but we can force it to do so ourselves by leveraging higher-order functions. Before discussing how eager versus lazy evaluation impacts the declarative code that we are writing, let's build a simple program that forces lazy evaluation in the preceding scenario. Once again, we'll create a list of numbers, filtered to only keep the even ones, and then pass them to numberPrinter:

```
func main() {
    input := []int{1, 2, 3, 4, 5, 6}
    isEven := func(i int) bool {
        return i%2 == 0
    }
    numberPrinter(func() []int {
        return Filter(input, isEven)
```

```
    })
    }
func numberPrinter(lazyGet func() []int) {
    fmt.Println("At this line, we don't yet know what our
        input values will be")
    for _, in := range lazyGet() {
        fmt.Println(in)
    }
}
```

In the preceding modified example, our numberPrinter function no longer takes a slice of integers as input. *Instead, it takes a function that returns a slice of integers as the input.* This is a crucial difference, as now it allows us to call the numberPrinter function without knowing the numbers to print ahead of time. Once numberPrinter deems it necessary to know the numbers, it can call the lazyGet function, which will generate each number. When we want to use numberPrinter, we have to provide a way for the function to get to the real input. We have done this with an anonymous function, simply wrapping our call to Filter in a new function that passes the output along:

```
    numberPrinter(func() []int {
        return pkg.Filter(input, isEven)
    })
```

This way, we can model lazy evaluation in Go. I'd say the main difference between this approach and what is typically considered a "lazily evaluated language" is that in a "lazy language," this type of laziness is handled by the programming language itself. In Go, while we get to lazily evaluate each intermediate result for a function, doing so would require a lot of overhead.

Delaying and avoiding execution

The right way to think about lazy evaluation is not simply delayed execution but rather *delay and avoid* execution. When working with lists, this translates into only generating the subset of the list required to solve the problem. This behavior is easy to mock in Go if we don't want to write declarative code and hand-write our loops, but it's much harder if we want to write declarative code. As mentioned earlier in the book, our aim is for code to be as declarative as possible, as this increases the readability. The next example will highlight what is meant by *delay and avoid* execution.

Let's say we want to find the first factorial result that is larger than 10 million, and we want to write this in a declarative way. To demonstrate this, we will also reuse what we learned in earlier chapters. We'll use the Maybe type introduced in *Chapter 5*, create a new function (head), attach this function to a slice type (ints), create a function to generate a pre-populated slice of integers (IntRange), and finally, tie this together into a single solution.

The complete example can be found on GitHub: `https://github.com/PacktPublishing/Functional-Programming-in-Go./tree/main/Chapter8/LazyEvaluation`. Let's start by setting up the `head` function:

```
func Head[A any] (input []A) Maybe[A] {
        if len(input) == 0 {
                return Nothing[A]()
        }
        return Just(input[0])
}
```

This function returns Maybe, which either contains the underlying head of the list, or returns `Nothing`. To attach this to a type to use in our dot notation chain, we'll need to provide a wrapper function:

```
func (i ints) Head() Maybe[int] {
        return Head(i)
}
```

Next, we need to generate a slice of numbers. The `IntRange` function will generate a range of numbers between a lower and upper bound. Remember that when writing declarative code, we want to concern ourselves with the *what* and not the *how*. As Go does not offer this out of the box, we'll write the generator function once (the *how*) and then only reuse the generator later (the *what*):

```
func IntRange(start, end int) []int {
        out := []int{}
        for i := start; i <= end; i++ {
                out = append(out, i)
        }
        return out
}
```

If we were to write enough of these types of generators, we would ideally never have to write a manual loop again. Now that we have written these functions, in combination with `Filter` from *Chapter 6* and `Factorial` from *Chapter 7*, we can tie this together into our declarative solution. In our `main` function, we will first create an inner function that checks whether a number is larger than 10 million. Then, we'll declaratively chain the steps to find the first factorial larger than 10 million by doing the following:

1. Generating a range from 0 to 100.

2. Mapping each number in the range to its factorial result.

3.   Filtering the results for being larger than 10 million.

4.   Returning the first element of this list:

```
func main() {
        largerThan10Mil := func(i int) bool {
                return i > 10_000_000
        }
        res := ints(IntRange(0, 100)).
                Map(Factorial).
                Filter(largerThan10Mil).
                Head()
        fmt.Printf("%v\n", res)
}
```

If we run this code, we get the following result – {39916800}.

While this is easy to read and understand, there is a pretty big drawback hiding underneath the implementation, due to Go's lack of lazy evaluation. What we have done in the first two steps is as follows:

1.   Generate all numbers from 0 to 100.

2.   Get their factorial result.

However, the first factorial that exceeded 10_000_000 actually happened at value for n=11. This means that every subsequent number (12 to 100) was generated and added to the slice, and then had its factorial calculated for no reason. In a lazily evaluated language such as Haskell, the list would only have generated the values needed to find the result and then short-circuited execution.

Short-circuiting in predicates

There is a form of short-circuiting and lazy evaluation that happens in most mainstream programming languages, including Go, which is short-circuiting for predicates. In an if condition, if there are multiple conditions present such as if A() && B(), the B function would not be executed if A already returned false. Similarly, for the if A() || B() statement, the B function would not be executed if A already returned true. This saves on useless computation. (If you are writing side effect-heavy code that would have relied on the result of running both predicates, this can be nasty to debug. Yet another reason to avoid side effects.)

Infinite data structures and lazy evaluation

Another advantage of lazy evaluation is that you can model infinite data structures, such as a list with all numbers from 0 to infinity. The reason that we can work with infinite structures in a lazily evaluated language is that you only compute as much data as is needed for the entire chain of operation. Go does not support lazy evaluation, so in this brief segue into the world of infinite data structures, the examples will be written in Haskell and an imaginary Go implementation.

In Haskell, defining an infinite list is a simple operation:

```
InfiniteInts :: [Int]
InfiniteInts = [1..]
```

So how do we work with them? Well, we need a terminating function. For lazy evaluation to work with infinite lists, we need to have a clear end state at which the list operation completes. For example, let's create an infinite list, check each number to see whether it is a prime number, and stop once we have generated 1 million prime numbers.

First, let's create the `naturals` function, which generates all numbers from 2 to infinity. The reason for doing so is that we don't know exactly where to stop. Let's also define what the sieve of Eratosthenes looks like:

```
naturals :: [Int]
naturals = [2..]

sieve :: [Int] -> [Int]
sieve (p:xs) = p : sieve [x | x <- xs, x `mod` p /= 0]
```

The sieve will remove (sieve out) all non-primes from the list for a given starting value. Next, let's wrap these two together into a function that generates prime numbers by feeding the infinite list of numbers into the sieve, alongside a limit (n) for how many we want to generate:

```
primes :: Int -> [Int]
primes n = take n (sieve naturals)
```

Here, we have our terminator function. `take` n tells us that, from the infinite list of data, we only want to generate however many are needed to reach n. Let's call this in a `main` function to generate the first 1 million of them:

```
main :: IO ()
main = do
  let millionPrimes = primes 1000000
```

```
putStrLn $ "Generated " ++ show (length millionPrimes)
    ++ " prime numbers"
```

The preceding code is all written in Haskell, but now let's move back to the domain of the star of this book, Go. If we think about how we would achieve something similar in Go, the easiest way to do so is by using a `for { }` loop. And to be specific, I mean the *while* behavior of the loop. We loop until a condition is met without a postcondition to increment a value. Ignoring the prime check, we would probably write something akin to the following:

```
func main() {
    primes := []int{}
    for len(primes) != 1_000_000 {
        // sieve or other algorithm to get prime
    }
}
```

The preceding implementation would work, assuming we fill out the body of the `for` loop (which is a de facto infinite generator; if we never reach the count of 1 million, it will keep looping forever. In practice, it means your algorithm is wrong). However, in writing this code, we have given up the declarative style of programming. *We moved back into the domain of spelling out "how" the result should be reached, rather than focusing on "what" the result should be.* In an imaginary Go implementation, what we would want to write is the following:

```
func main() {
    millionPrimes :=
        IntRange(2
            Filter(func(i int) bool {
                return isPrime(i)
            }).
            Take(1_000_000)
}
```

The preceding code would be the equivalent (functional) implementation, although we are filtering instead of using a sieve for the sake of simplicity. This ends our segue into lazy evaluation and its benefits. Let's move on to another style of chaining functions together.

Continuation-passing style programming

The next programming style that we will look at is **continuation-passing style** (**CPS**). Unlike the familiar dot notation style of method chaining, CPS is only possible in languages that support functions as first-class citizens. The core idea is that the continuation – in other words, the next step of execution

– is another function that is passed as an argument to our original function. This allows us to control the flow of our program using function passing, rather than by branching and explicit function calls. The main benefit is that this will help us read and understand complex chains of functions, and we can change them with minimal effort. Before we dive into the Go implementation of CPS programming, let's take a brief detour and explain continuations.

What are continuations?

A continuation is a somewhat abstract concept in the realm of programming languages. It is a function that represents the next computation of a program. It essentially captures the state of our program at the moment of execution (more specifically, the stack), and provides the next step of execution as a function that can be called.

Continuations are used to implement the control flow within a programming language. They can be thought of as a data structure that represents our current state of execution and the next state of execution that we will transition into. This abstract concept is how programming languages can implement the control flow structures that are more familiar to us from day-to-day programming, such as exception handling, `for` loops, and goroutines.

In some languages, such as Scheme, continuations are exposed to the programmer and can be leveraged to control the flow of execution of the program at a higher level of abstraction. This is equivalent to programming your own control structs but with the additional advantage that the continuation can be modified in place to take on different behavior. In Go, this is not easily achievable. One of the challenges in doing this in Go is that it is a statically typed language, making it harder to even begin to define a continuation as a data structure.

The closest example that touches on continuations in Go is perhaps the `panic` and `recover` pattern. Imagine the following function:

```go
func main() {
    defer func() {
        if r := recover(); r != nil {
            fmt.Println("Recovered from panic:", r)
        }
    }()
    fmt.Println("Normal execution happening")
    panic("Execution flow is broken")
    fmt.Println("This line will not be executed")
}
```

In this `main` function, we are first defining a **deferred** function that runs at the end of the `main` function, just before function exits. `defer` specifies a continuation for a function and is a special case in Go, in that it gets executed prior to function exit, regardless of how we exit the function. Inside the deferred function, we will recover if `panic` was encountered anywhere during the execution of the parent function.

Outside of the deferred function, we are explicitly calling panic after the first print statement. This is, again, an example of a continuation. There are a few special things happening here that might not be immediately obvious. First, the call to panic is not a simple function call. panic is used to signal that our execution stack ended up in a corrupted state, and the normal flow of execution is no longer possible. It captures the entire state of the stack at the moment panic was called and halts our function. However, because we have added the defer function as a continuation to main, defer gets access to this saved stack from panic. This can then display the content of the stack at the moment that panic was called, help us recover from the invalid state, and continue program execution without halting. In other words, our panic continuation is capturing valuable information about the state of our program and is exposing this to another function later on during execution. However, do remember that the use of panic is discouraged in Go.

Diving in depth into continuations is beyond the scope of this book, but this small introduction should serve the purpose of showing that continuations, although not explicitly called such in Go, are still prevalent in the language itself. There is a form of explicit continuation that we can leverage though, and that is how we end up in the domain of CPS programming.

Implementing CPS code in Go

To build an understanding of CPS, let's first take a look at a simple example. Recall from the last chapter that we created a few ways to calculate the factorial of a number. In this example, let's rewrite the recursive version to follow the CPS pattern. To enable CPS, we need to pass the continuation as an argument to our recursive function. The remainder of the logic for calculating the factorial remains the same:

```
func factorial(n int, f func(int)) {
    if n == 1 {
        f(1) // base-case
    } else {
        factorial(n-1, func(y int) {
            f(n * y)
        })
    }
}
```

In this factorial example, we are using a higher-order function, f, to represent the continuation of our recursive function call. Whereas in the recursive function call, we simply multiply the current input of our function by the result of the function call, here we are using a closure to perform the multiplication one layer deeper. We are saying that to continue calculating the factorial, we need to multiply n by y. However, y is not yet defined in this stack frame; it will only be defined in the next function call.

To run this function, let's create a main function that prints the result of the multiplications:

```
func main() {
    factorial(5, func(i int) {
        fmt.Printf("result: %v", i)
    })
}
```

Note how, in this function, we are calling `fmt.Printf` inside the closure. This means that the print statement will get passed down our function call chain, and it will eventually be evaluated by our factorial function. This is one of the powers of CPS – it is making explicit what happens at lower steps of recursion, rather than this being hidden from the programmer. The topmost stack frame gets a function pushed to it, which is the `print` function, and each subsequent call stack gets a function pushed onto it, which is a "multiply our input with the input of the next call" function.

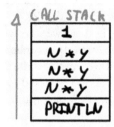

Figure 8.4: Call stack with variable functions

If we take a look at *Figure 8.4*, we can see how there are different functions being pushed to different levels of our call stack. The bottommost call stack has a function call to `Println`, while the ones immediately above it have a function call to a multiplication closure. The final stack frame simply has the constant **1** (to signify n * 1). By introducing CPS, we have effectively abstracted over the recursive function call and have added additional flexibility to how our program flows through the call stack. Whereas in a normal recursive function, each subsequent layer is mostly a copy of the previous layers (apart from the base case), with CPS, we can introduce a different behavior at each frame, depending on what we want to do.

One thing to point out when we are using this mental model of call frames is that they are never returned. In a CPS program, you are not using `return` statements; rather, you are passing the continuation (next step) as a higher-order function. This is why our `print` statement is passed down until the final iteration is reached. To be explicit about it, each call frame is added to the stack but evaluation rolls up from the bottom – that is, our `print` statement is pushed to the frame above it, which pushes it to the frame above it, and so on, all the way to the top.

The function result is rolled up along the way with it. This is in contrast with normal recursion, where our stack frames are added similarly but evaluated from top to bottom. It can take a second to wrap your head around this inversion of evaluation flow. The reason behind this is that in each stack frame, we are passing closures as *input* to the next function. *However, remember that passing around a function as input to another function does not evaluate that function yet.* Hence, we are delaying the execution (as with lazy evaluation) of each function along the stack frame until the final frame is reached. In this case, `f(1)` is the final stack frame. Once we have reached this frame, all our closure functions are effectively evaluated. (From the last closure to the first closure, they are evaluated in the final frame. Hence, the `print` statement from the bottom prints the result of the final evaluation.)

Now that we have seen this example work with recursion, let's look at a slightly more complex example where we are not actually using recursion. This is to show that any type of control flow can actually be modeled with CPS.

Simple mathematics operations with CPS

In the preceding example, we have seen the recursive factorial calculation using CPS. This might lead us to believe that CPS is just an alternative way of writing a recursive function. And while there are certain advantages to writing recursive functions with CPS, it is not strictly necessary for a function to be recursive. Take the following example. Let's say we start with a slice of integers as input. We first want to filter out the even numbers from the input. If a number is even, we want to double it. Finally, we want to print out the resulting integer. If we want to write this in CPS, we need to consider each continuation (following action) as a function to be passed to the original function. This, without diving into the details, produces the following flow of continuations:

```
Input []int -> isEven(int) -> double(int) -> print(int)
```

This shows us that three continuation functions are needed, along with a fourth function in which we will create the slice of inputs and start the chain of operations. Written in Go, this would produce the following result:

```
func main() {
    is := []int{1, 2, 3, 4, 5, 6}
    isEven(is, func(i int) {
        double(i, print)
    })
}
func isEven(input []int, cont func(int)) {
    for _, i := range input {
        if i%2 == 0 {
            cont(i)
        }
    }
}
```

```
}
func double(input int, cont func(int)) {
    cont(input * 2)
}
func print(i int) {
    fmt.Println(i)
}
```

Each of our functions, apart from `print`, performs an operation on our input and calls a continuation function. The continuation function will provide the next step in our algorithm. In the `isEven` function, the continuation is only called if a number matches the `i%2==0` condition, effectively making sure that the continuation only happens on numbers that are even. Now, when we read our `main` function, the entire chain of operations is spelled out:

```
func main() {
    is := []int{1, 2, 3, 4, 5, 6}
    isEven(is, func(i int) {
        double(i, print)
    })
}
```

First, we create a slice of integers. Next, we call `isEven`; after that, we double, and finally, we print. Note that the odd thing here is that we are actually creating an anonymous function as input for the `isEven` continuation. In Go, we cannot simply write the function like this:

```
func main() {
    is := []int{1, 2, 3, 4, 5, 6}
    isEven(is,    double(i, print))
}
```

Well, we could if we changed the signature of `isEven` to take as input arguments:

```
func isEven(input []int, cont func(int, func(int))) {
```

However, now our `isEven` function is tied to a function that explicitly takes a continuation as a request. What if we simply want to print the even numbers and perform no further operations on them?

This gets to the heart of why CPS is difficult to get right in Go. The type system is too strict to easily manage functions in the CPS style of programming. We will discuss the drawbacks of CPS in more detail, but first, let's take a look at a scenario where CPS can actually be a real advantage.

CPS and goroutines

One of the areas where CPS can definitely help is managing concurrent code. Whenever you hear about a *callback* in languages such as JavaScript, what you are really doing is using a continuation and passing this to an asynchronous function. Once the asynchronous part has been completed, the continuation (callback) is automatically called with the result of the asynchronous part. Often, this is done in the form of web requests, whereby a web request is launched, the callback is called when the request has been completed, and the state of the callback is populated with the result of the request. This result is often a status code (e.g., 200) and a payload (in the case of a GET request). This is such a common pattern nowadays that we ignore the underlying concepts, nor do we really need to understand them to use callbacks. However, let's model our *double if even* function with callbacks and asynchronous Go code to provide an explicit example:

```
func main() {
        callback := func(input int, b bool) {
                if b {
                        fmt.Printf("the number %v is
                                even\n", input)
                } else {
                        fmt.Printf("the number %v is
                                odd\n", input)
                }
        }
        for i := 0; i < 10; i++ {
                go isEven(i, callback)
        }
        _ := <-make(chan int)
}
func isEven(i int, callback func(int, bool)) {
        if i%2 == 0 {
                callback(i, true)
        } else {
                callback(i, false)
        }
}
```

The preceding code is a slightly altered version. We are going to asynchronously verify whether or not a number is even, and if it is, we will print the number x is even; otherwise, we will print the number x is odd. The key component that CPS aims to simplify here is the flow-control part of an asynchronous call. We will launch a call with the go keyword, and as the continuation is encoded as part of the function that we are calling, we don't have to worry about waiting for the result of the function call asynchronously to then launch into the next function. The pattern of *launch call, wait for result, continue computation* is often modeled as async/await operations in languages that don't support higher-order functions. As Go is a multi-paradigm language, we can leverage higher-order functions and, thus, CPS. This allows us to focus on the *async* part without having to worry about the *wait and continue* part. That said, Go actually has a solid concurrency paradigm with goroutines and channels, so the need for this CPS-style programming is mostly alleviated.

When to use CPS?

For most use cases, CPS will likely make your program more complex than it's worth. It's not the easiest way of reading recursive functions, and even when you are used to it, it can trip you up. It is, however, used in certain spaces, such as compiler/interpreter design. Typically, if there is a complex control flow that you want to model, CPS can make this control explicit and, thus, easier to comprehend and read.

Apart from this, another use case is the use of callbacks in asynchronous programming. Although we don't often call those CPS, or even continuations, they are definitely a form of CPS. Due to the use of goroutines and channels, the more familiar style of callbacks that we find in languages such as JavaScript is a bit less common but, nevertheless, a useful area where we can use them.

Summary

In this chapter, we have looked at two distinct ways of composing our functional code. The first way is by chaining methods in a familiar dot notation-style chaining. This is a way of connecting inputs and outputs of various functions without having an intermediate variable assignment in between. While most programmers are familiar with this style of programming, there is some overhead required when writing (pure) functional code with generics in Go.

Another trade-off that we discussed here is the eager versus lazy modes of function evaluation. While it is possible to mimic lazy evaluation in Go, the compiler and language don't do any of the heavy lifting for us. This means that if we were to port code from a functional language such as Haskell, the performance characters would be significantly different.

Finally, we also looked at continuations and CPS programming. A continuation is an abstract representation of any "next step" in an algorithm, whether it is a function call, a loop, or a "goto" statement. CPS programming makes the nature of recursive operations explicit and allows us to abstract over how function chaining happens. While CPS is a powerful technique, the use cases in daily life are a bit limited, even though we are using a lot of CPS under the hood, such as when modeling callback functions.

In the next chapter, we will jump one layer of abstraction higher and look at program composition through functional design patterns.

Part 3: Design Patterns and Functional Programming Libraries

In this part, we will first move to a higher level of abstraction by looking at software architecture using functional programming techniques. Once again, we will compare how the object-oriented approach compares with the more functional approach. We'll see how Go's concurrency paradigm can be leveraged in a functional context. Finally, we will learn about libraries that help us in building functional applications.

This part has the following chapters:

- *Chapter 9, Functional Design Patterns*
- *Chapter 10, Concurrency and Functional Programming*
- *Chapter 11, Functional Programming Libraries*

9

Functional Design Patterns

In this chapter, we will move to a higher level of abstraction. Rather than talking about individual functions and operations, let's take a look at some design patterns. While we will not extensively explain each design pattern, we will take a look at how the object-oriented pattern translates to the functional world.

In this chapter, we're going to cover the following main topics:

- Classical design patterns in a functional paradigm:

  - The strategy pattern

  - The decorator pattern

  - The Hollywood principle

- Functional design patterns

Technical requirements

In this chapter, any version at or above Go 1.18 will work for all Go-related code. Some snippets are written in Java; those will work with any version of Java above 1.5.

The code for this chapter can be found on GitHub at `https://github.com/PacktPublishing/Functional-Programming-in-Go./tree/main/Chapter9`.

Classical design patterns in a functional paradigm

Anyone who has programmed in an object-oriented language will encounter design patterns at some point. Design patterns are a type of cookie-cutter solution to common engineering problems. One key point is that the solution they provide should be thought of as a starting point, a way to tackle a problem that has proven itself to be useful. Often, the solution is not readily usable out of the box and needs to be adapted to your concrete environment and situation. A given design pattern might provide 90% of a solution to a problem, and the remaining 10% is filled in with custom, non-pattern code.

This chapter does not aim to exhaustively cover design patterns. In fact, entire books have been written about design patterns, such as the well-known *Gang of Four* book, *Design Patterns: Elements of Reusable Object-Oriented Software*. What this chapter does aim to do is to showcase how certain object-oriented design patterns translate to the functional paradigm, and how they are often simpler to express in this paradigm. For each design pattern, we will take a look at the object-oriented implementation, the general problem and benefit of the pattern, and finally, what the functional implementation looks like. We'll start off with the strategy pattern and continue with the decorator pattern and the **Inversion of Control** (**IoC**) principle.

These are three patterns that are common to object-oriented code. The strategy pattern is a way to change the behavior of our program at runtime and decouple a class with a concrete implementation. The decorator pattern allows us to dynamically extend functions without breaking the open-closed principle, and the IoC principle is a staple of many object-oriented frameworks, whereby the order of control is delegated to the highest level in the call tree.

The strategy pattern

The first pattern that we will take a look at is the strategy pattern. The strategy pattern is a design pattern that allows us to dynamically change the algorithm of a method or function at runtime. By doing this, we can modify the behavior of our program throughout its runtime. In the example that we will work out, we will have an `EncryptionService`, which supports various ciphers.

We'll keep it simple and work with substitution ciphers that change the letters in the output. We will implement three different cipher mechanisms:

- The Caesar cipher
- The Atbash cipher
- A custom cipher

Each cipher needs to support the encryption and decryption of a given string as follows:

```
Input = decipher(cipher(Input))
```

In other words, we should be able to reconstruct the input from a ciphered output. For our implementations, we will also limit ourselves to changing the letters of the alphabet a-z, and ignore casing.

> **Ciphers and security**
>
> It bears calling out that these ciphers should never be used for actual encryption. They are incredibly weak and offer no real protection against a malicious actor in this day and age. They are interesting to study for their historical context and are fun to implement while being easy to understand.

Object-oriented strategy pattern

First, we will solve this problem in an object-oriented way. Remember that Go is a multi-paradigm language, so we can easily apply object-oriented design patterns in Go. *Figure 9.1* shows the architecture for this solution:

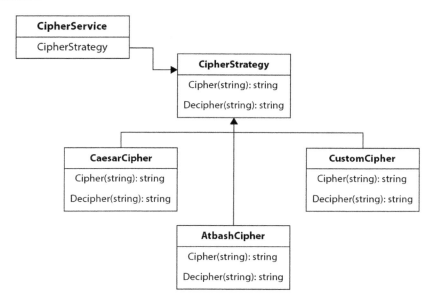

Figure 9.1: Strategy pattern for cipher implementation

In the object-oriented implementation, we start with a `CipherService`. This is any class that wants to use a cipher. Rather than having a concrete implementation, `CipherService` instead contains a `CipherStrategy` through object composition. This `CipherStrategy` is an interface that specifies the `Cipher` and `Decipher` methods. Both methods take a string as input and return either a ciphered or deciphered string. In *Figure 9.1*, we have three concrete implementations for the cipher:

- Caesar

- Atbash

- A custom cipher

Each of these is a class (struct) that implements the required methods (`Cipher` and `Decipher`). We can also include a useful state in these classes, as we will see in the upcoming code examples, whereby we maintain a `Rotation` variable as part of the Caesar cipher. Both the Caesar cipher and the Atbash cipher are so-called substitution ciphers. They exchange one letter of the alphabet with another letter. In the case of the Caesar cipher, the replacement letter is situated a certain amount of positions further in the alphabet. For the Atbash cipher, this is a simple substitution of each letter with the letter of the same position in the reverse alphabet (z-a).

Caesar

Let's start implementing this in Go. First, we'll set up the `CipherService`, as well as a slice containing all letters of the alphabet that we'll support. We will also need to figure out the index of a given rune in this slice of letters, which we will do by implementing an `indexOf` function:

```go
var (
    alphabet [26]rune = [26]rune{'a', 'b', 'c', 'd', 'e',
    'f', 'g', 'h', 'i', 'j', 'k', 'l', 'm', 'n', 'o', 'p',
    'q', 'r', 's', 't', 'u', 'v', 'w', 'x', 'y', 'z'}
)
func indexOf(r rune, rs [26]rune) (int, bool) {
    for i := 0; i < len(rs); i++ {
        if r == rs[i] {
            return i, true
        }
    }
    return -1, false
}
type CipherService struct {
    Strategy CipherStrategy
}
```

To follow a more traditional object-oriented language pattern, we can attach a `Cipher` and `Decipher` method to `CipherService` as well. This will just delegate the call to the chosen implementation (`Strategy`):

```go
func (c CipherService) Cipher(input string) string {
        return c.Strategy.Cipher(input)
}
func (c CipherService) Decipher(input string) string {
        return c.Strategy.Decipher(input)
}
```

After this is set up, we will also define an interface, `CipherStrategy`, which will enforce any implementation to have the `Cipher` and `Decipher` methods:

```go
type CipherStrategy interface {
    Cipher(string) string
    Decipher(string) string
}
```

With this in place, we can start implementing the strategies that we will support. For brevity, we will only implement the Caesar and Atbash cipher. Implementing a custom cipher, as in *Figure 9.1*, would be a trivial extension of this. To implement the Caesar cipher, we will first define a struct to represent this strategy:

```
type CaesarCipher struct {
    Rotation int
}
```

The Caesar cipher is a cipher whereby a letter in the input is exchanged for a letter a certain number of positions further in the alphabet. The number of positions that we use is defined as the *rotation* of the cipher. For example, if we have the abc input and a rotation of 1, each letter is replaced with the letter 1 position further in the alphabet, so the output would be bcd.

Similarly, if the rotation were 2, the output would be cde, and so on. The following is an implementation of the Caesar `Cipher` and `Decipher` methods in Go. Understanding the implementation is not that important; the important part is to note how we select which implementation `CipherService` uses and even change it during the execution of our program:

```
func (c CaesarCipher) Cipher(input string) string {
    output := ""
    for _, r := range input {
        if idx, ok := indexOf(r, alphabet); ok {
            idx += c.Rotation
            idx = idx % 26
            output += string(alphabet[idx])
        } else {
            output += string(r)
        }
    }
    return output
}
func (c CaesarCipher) Decipher(input string) string {
    output := ""
    for _, r := range input {
        if idx, ok := indexOf(r, alphabet); ok {
            idx += (26 - c.Rotation)
            idx = idx % 26
            output += string(alphabet[idx])
```

```
                    } else {
                            output += string(r)
                    }
            }
            return output
    }
```

Now that we have the Caesar cipher implemented, let's also implement the Atbash cipher.

Atbash

The Atbash cipher is a straightforward replacement of each letter with the letter at the same index but with the alphabet in reverse. So, a becomes z, b becomes y, and on until z becomes a. As a result, deciphering can be achieved by calling the cipher again, as we are effectively mirroring the alphabet (and mirroring twice returns the original result).

We don't need any real state to manage with the AtbashCipher struct, unlike CaesarCipher, where we maintained the rotation as a class variable. However, we will still need to create the struct for our strategy pattern implementation to work correctly. It will just be an empty struct with functions attached to it:

```
type AtbashCipher struct {}
func (a AtbashCipher) Cipher(input string) string {
        output := ""
        for _, r := range input {
                if idx, ok := indexOf(r, alphabet); ok {
                        idx = 25 - idx
                        output += string(alphabet[idx])
                } else {
                        output += string(r)
                }
        }
        return output
}
func (a AtbashCipher) Decipher(input string) string {
        return a.Cipher(input)
}
```

Again, the actual implementation of the code here is not that important. It is neat that we can decipher it by just calling `Cipher` again, and this will become even more interesting in the functional example. Either way, let's look at how we can change the implementation during execution and switch between `CaesarCipher` and `AtbashCipher`:

```
func main() {
        svc := CipherService{}
        svc.Strategy = CaesarCipher{Rotation: 10}
        fmt.Println(svc.Cipher("helloworld"))
        svc.Strategy = AtbashCipher{}
        fmt.Println(svc.Cipher("helloworld"))
}
```

This is the object-oriented implementation of the strategy pattern. We have created three classes (`CipherService`, `CaesarCipher`, and `AtbashCipher`) one interface (`CipherStrategy`), and two functions per struct (to cipher and decipher). Now, let's take a look at a functional implementation.

Functional implementation of the strategy pattern

We have already seen in previous chapters how we can dynamically change the implementation details of an algorithm by leveraging the fact that functions are first-class citizens, and we can pass them around like objects in a traditional object-oriented language. If we refactored our `CipherService`, all we would need to know is that this service needs a function to take a string and return a string twice (one for ciphering and one for deciphering).

To start off, let's define the struct for this new service, as well as two types to define the `Cipher` and `Decipher` functions:

```
type (
        CipherFunc    func(string) string
        DecipherFunc func(string) string
)
type CipherService struct {
    CipherFn    CipherFunc
    DecipherFn DecipherFunc
}
func (c CipherService) Cipher(input string) string {
    return c.CipherFn(input)
}
func (c CipherService) Decipher(input string) string {
```

```
        return c.DecipherFn(input)
}
```

Now that we have `CipherService` in place, we need to define our Caesar and Atbash cipher-related functions. Unlike in the object-oriented example, we don't need to define a new struct to do so. We can define our functions in the same package as our `CipherService` but we would not have to do so. In fact, any function of the correct type can be used as a `Cipher` or `Decipher` function.

Let's implement `CaesarCipher` first. The one thing we do have to be aware of is that we do not have a struct that can hold the state anymore. In our example, the `CaesarCipher` struct stored `Rotation` as a class variable. In the functional approach, the rotation needs to be part of the `CaesarCipher` function itself. It's a minor but important change. Apart from this change, the implementation remains the same:

```
func CaesarCipher(input string, rotation int) string {
    output := ""
    for _, r := range input {
        idx := indexOf(r, alphabet)
        idx += rotation
        idx = idx % 26
        output += string(alphabet[idx])
    }
    return output
}
func CaesarDecipher(input string, rotation int) string {
    output := ""
    for _, r := range input {
        idx := indexOf(r, alphabet)
        idx += (26 - rotation)
        idx = idx % 26
        output += string(alphabet[idx])
    }
    return output
}
```

Similarly, we can implement `AtbashCipher` as a function. One nice thing here is that due to the relationship between ciphering and deciphering with Atbash, we don't have to actually write any implementation for the `Decipher` function. Rather, we can just equate the `Decipher` function to the `Cipher` function:

```
func AtbashCipher(input string) string {
    output := ""
    for _, r := range input {
        if idx, ok := indexOf(r, alphabet); ok {
            idx = 25 - idx
            output += string(alphabet[idx])
        } else {
            output += string(r)
        }
    }
    return output
}
var AtbashDecipher = AtbashCipher
```

The last line effectively defines a new function, `AtbashDecipher`, with the same implementation as `AtbashCipher`, once again leveraging the fact that our functions are simply data, which can be stored as variables in Go.

When using this functional implementation in Go, we have to provide a function of the `func (string) string` type to both the `Cipher` and `Decipher` implementation of our service. As a result of `CaesarCipher` requiring an extra variable to determine the rotation, we do need to create a closure for our `CipherService`. In our `main` method, we can dynamically update the cipher that we want to use to `AtbashCipher` without the need for a closure, as the Atbash cipher is a straightforward cipher that adheres to `func (string) string`:

```
func main() {
    fpSvc := {
        CipherFn: func(input string) string {
            return (input, 10)
        },
        DecipherFn: func(input string) string {
            Return fp.CaesarDecipher(input, 10)
        },
    }
    fmt.Println(fpSvc.Cipher("helloworld"))
```

```
        fpSvc.CipherFn = AtbashCipher
        fpSvc.DecipherFn = AtbashDeciphe
        fmt.Println(fpSvc.Cipher("helloworld"))
        fmt.Println(fpSvc.Decipher(fpSvc.Cipher("hello")))
    }
```

This example prints some ciphered and deciphered content using our functional implementation. Using this functional implementation, we could easily implement ad hoc ciphers without defining them as standalone functions. Both the `Cipher` and `Decipher` implementation accept anonymous functions to specify the implementation details. This is what we have done to make the Caesar cipher work by wrapping it in such an anonymous function.

The decorator pattern

Let's modify our code to also adhere to the decorator pattern. The decorator pattern is a way to add functionality to our methods and classes without having to modify them. This means that the *open-closed* part of SOLID is respected. When programming in an object-oriented fashion, this is done through function composition (and often with inheritance in languages that support this). In Go, composition is the favored way of composing structs, so the decorator pattern feels natural for both a functional and object-oriented style implementation.

> **SOLID principles for object-oriented design**
>
> SOLID is a set of principles for designing robust object-oriented systems. It stands for **Single-Responsibility, Open-Closed, Liskov Substitution, Interface Segregation, and Dependency Inversion**. These principles are good to adhere to regardless of which paradigm you use, but their implementation differs. For example, functions should have a single responsibility, be closed to modification but open to extension, and functions should rely on abstract (higher-order) functions rather than concrete implementations.

Object-oriented decorator pattern

First, let's start off by implementing the decorator pattern in an object-oriented way. We'll extend our strategy pattern example of the various ciphers. To keep things simple, let's just say we want to log the input to each `Cipher` and `Decipher` function. To make our program more composable, we don't want to add the `log` function by modifying the existing `CaesarCipher` and `AtbashCipher` structs. If we were to do so, we would also have to update the `log` functionality for each struct in case the logging requirements change. Instead, what we will do is implement a `LogCipherDecorator` struct. This struct adheres to the `CipherStrategy` interface by implementing a function for both `Cipher` and `Decipher`. These functions will first write to a log and then delegate each call to the underlying `Cipher` or `Decipher` implementation. *Figure 9.2* shows the class diagram for this pattern.

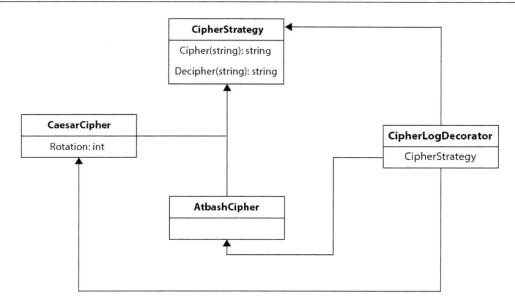

Figure 9.2: Class diagram for the decorator pattern

Now, we can turn this into code; let's look at the struct definition first. We have a new `LogCipherDecorator` struct, which uses `CipherStrategy` through composition:

```
type CipherLogDecorator struct {
    CipherI CipherStrategy
}
```

Now, we will also implement the necessary functions to have this new struct adhere to `CipherStrategy` itself. In each function, first, we will log the input prior to dispatching the call to the underlying `CipherStrategy`:

```
func (c CipherLogDecorator) Cipher(input string) string {
    log.Printf("ciphering: %s\n", input)
    return c.CipherI.Cipher(input)
}
func (c CipherLogDecorator) Decipher(input string) string {
    log.Printf("deciphering: %s\n", input)
    return c.CipherI.Decipher(input)
}
```

That is essentially all that is required to implement the decorator pattern. It comes in handy in a variety of scenarios, but it's encountered especially often when working with **User Interface** (**UI**) code (Java UI libraries such as Swing tend to use this extensively).

In the `main` function, we can now use `CipherLogDecorator` anywhere that we expect `CipherStrategy`. We'll have to instantiate the decorator with the underlying class to get the additional functionality:

```
func main() {
    cld := {
        CipherI: oop.CaesarCipher{Rotation: 10},
    }
    svc := oop.CipherService{Strategy: cld}
    ciphered := svc.Cipher("helloworld")
    fmt.Println(ciphered)
}
```

In this snippet, we can see how `CipherService` accepts `CipherLogDecorator` just like any other `CipherService`. When we run this `main` function, the log statement appears before each print statement. Running that function, we get the following:

```
[ec2-user@ip-172-31-29-49 Chapter9]$ go run main.go
2023/01/14 15:50:05 ciphering: helloworld
rovvygybvn
```

Next, let's functionally implement this and compare the two approaches.

Functional decorator pattern implementation

Applying the decorator pattern to functional programming does not require anything that we haven't seen before in this book. We have learned about function composition and used that in previous chapters. The decorator pattern for object-oriented code really is nothing more than function composition in the functional programming paradigm.

As such, creating a function for adding a log statement prior to each `cipher` or `decipher` call is a matter of creating a higher-order function that takes either a `Cipher` or `Decipher` function as input, and returns a new function, which first calls `log` and then delegates the remainder of the functionality to the underlying function. Let's make this concrete by looking at the decorating functions for ciphering and deciphering, `LogCipher` and `LogDecipher`:

```
func LogCipher(cipher CipherFunc) CipherFunc {
    return func(input string) string {
```

```
        log.Printf("ciphering: %s\n", input)
        return cipher(input)
    }
}
func LogDecipher(decipher DecipherFunc) DecipherFunc {
    return func(input string) string {
        log.Printf("deciphering: %s\n", input)
        return decipher(input)
    }
}
```

In essence, that's all that needs to happen to decorate functions with new functionality. LogCipher accepts any function that adheres to the CipherFunc type definition and returns a new function that also adheres to that type definition. The new function, created as an anonymous function returned from LogCipher, then calls log and subsequently our CipherFunc that was passed initially.

The main difference in the implementation strategy with the object-oriented and functional paradigms is just how we define adherence to the expected functionality. With an object-oriented approach, we use interfaces to define adherence, while with the functional approach, we use the type system to define adherence.

In our main function, we can create CipherService using the decorator functions rather than the underlying ciphers:

```
func main() {
    caesarCipher := func(input string) string {
        return CaesarCipher(input, 10)
    }
    caesarDecipher := func(input string) string {
        return CaesarDecipher(input, 10)
    }
    fpSvc := {
        CipherFn:   LogCipher(caesarCipher),
        DecipherFn: LogDecipher(caesarDecipher),
    }
    fmt.Println(fpSvc.Cipher("hello"))
}
```

Notice that in this example, for readability, the decorator functions are split from the creation of `CipherService`, but this could have been done in line, as in the earlier strategy pattern implementation. If we created `CipherService` with `AtbashCipher` instead, it makes for a more readable example:

```
func main() {
    fpSvc := fp.CipherService{
        CipherFn:   fp.LogCipher(caesarCipher),
        DecipherFn: fp.LogDecipher(caesarDecipher),
    }
    fmt.Println(fpSvc.Cipher("hello"))
}
```

As we can see from the examples, function composition is the key to decorating functions with additional functionality, which can then be shared across implementations. Another advantage of what we have done so far can be described as the *Hollywood principle*, also known as the *IoC* principle.

The Hollywood principle

The Hollywood principle of *don't call us, we'll call you* is also known as the IoC principle. IoC is an abstraction of the well-known Dependency Injection pattern. Dependency Injection is an important aspect of writing object-oriented applications and is useful for the functional paradigm as well.

Without going too in-depth into the object-oriented implementation, the key takeaway is that objects should defer concrete implementations of their dependencies to the highest level in the object/call hierarchy. We have done this implicitly in the previous examples for our cipher implementations by leveraging interfaces rather than concrete implementations. Notice that the object-oriented `CipherService` did not specify which cipher it would use and instead, we deferred that choice to the creator of `CipherService` by just asking for an implementation of the `CipherStrategy` interface:

```
type CipherStrategy interface {
    Cipher(string) string
    Decipher(string) string
}
type CipherService struct {
    Strategy CipherStrategy
}
```

Go lends itself quite naturally to this way of programming by not having explicit constructors for structs. In a language such as Java, where objects can be instantiated with default class-level objects through object composition, it is easier to ignore programming against an abstract implementation.

For example, the following Java snippet would show an implementation of a `CipherService` that does not adhere to IoC but uses a specific type of cipher (the Caesar cipher, in this case):

```java
class CaesarCipher {
    int rotation;
    CaesarCipher(int rotation) {
        this.rotation = rotation;
    }
}
class CipherService {
    CaesarCipher cipher = new CaesarCipher();
    CipherService() {}
    public String cipher(String input) {
        String result = "";
        // implement cipher
        return result;
    }
}
```

Why do we highlight this Java code here? First, to show that Go's struct paradigm lends itself naturally to IoC by way of struct instantiation without constructors. This means that structs do not have an inherent class state.

This brings us to the functional implementations of services. In Go, we have two ways of making IoC happen:

- The first way is through the use of interfaces, as we have done in the object-oriented examples
- The second way is to use type definitions and functions as first-class citizens to abstract over the behavior of a struct

To illustrate the difference, the following are the two definitions of the `CipherService` that we have used, and both apply IoC in alignment with their paradigm.

First, let's show the object-oriented way:

```go
type CipherStrategy interface {
    Cipher(string) string
    Decipher(string) string
}
type CipherService struct {
```

```
    Strategy CipherStrategy
}
```

And now the functional way:

```
type (
    CipherFunc    func(string) string
    DecipherFunc func(string) string
)
type CipherService struct {
    CipherFn    CipherFunc
    DecipherFn DecipherFunc
}
```

This was just a brief segue to point out what is happening in both instances. Let's continue our discussion of design patterns.

Functional design patterns

In the preceding sections of this chapter, we have compared functional and object-oriented design patterns (strategy, decorator, and Dependency Injection/IoC). If we look at the main differences between the functional and object-oriented patterns, it becomes clear that our patterns are achieved through different combinations of functions. We are either using functions as first-class citizens to store them as variables within a struct, or we are using function composition, higher-order functions, anonymous functions, and closures to achieve what would traditionally have been achieved with interfaces and the inheritance of classes.

And this really should be the main takeaway when writing functional code. Everything is a function. Design patterns become patterns of function combinations. As such, there is no real counterpart to the traditional design patterns outlined by the *Gang of Four* for the object-oriented world. So, what does the functional paradigm offer in terms of design patterns? Well, if we go back to the definition of design patterns, we can see that a pattern is a reusable solution to a commonly encountered problem. It is a cookie-cutter approach that might solve 85% of your problem, while the remaining 15% remains to be solved beyond the pattern. Functional programming does offer these solutions, and we discussed many of them earlier in this book.

When you think of function currying to compose different functions together and reducing each function to a 1-ary function to then combine them into any n-ary function, these steps can be thought of as a functional design pattern. Similarly, using closures, monads, and callbacks through CPS all can be thought of as patterns that are applied to solve a common problem. What we don't have in functional programming is the overhead of object taxonomy, which is what the design patterns in object-oriented code reflect. You could argue that the need for design patterns in traditional object-oriented languages is more of a solution to a limitation in the programming language itself rather than a real benefit to the programmer.

One of the ways traditional design patterns are avoided is through the use of function composition, but an equally critical component is leveraging the type system – a type system that can assign a concrete type to a function of a specified signature. Looking at object-oriented design patterns, whether the decorator pattern, factory pattern, or visitor pattern, they leverage interfaces extensively to abstract the implementation details. In Go, we can use the type system to abstract the implementation, as we have done in the earlier examples.

If we summed up how to solve a particular design problem in the functional paradigm, it would be rather boring, as the problem either does not exist or is solved through functions. Our solution would just look like *Table 9.1*:

Design Pattern	Solution
Strategy pattern	Functions (higher-order functions + function types)
Decorator pattern	Function composition (closures)
Factory pattern	No real need, as we don't need objects, but we could create functions with a set of default values – so, this would be function currying
Visitor pattern	Functions
Singleton pattern	No need, as we avoid objects and mutable state
Adapter	Could be seen as function mapping
Façade	Functions again

Table 9.1: Design patterns and their functional solution

In Go, however, we are working with a multi-paradigm language, so we get to have the best of both worlds. We can leverage some design patterns when we are working with structs, but their implementation is simplified in many ways through the use of functional programming principles rather than object-oriented ones. Despite creating an interface to abstract the implementation of a struct's functionality, we can use a function that adheres to a given type, as we have done with `CipherService`.

Summary

In this chapter, we took a look at the design patterns common in object-oriented code, namely the strategy pattern, the decorator pattern, and the Hollywood principle (IoC). We saw that these can be implemented in Go without the need for extensive object taxonomy simply by leveraging functions as first-class citizens. We also discussed the need for design patterns in the functional paradigm and concluded that either the patterns are not needed or can be solved using functions. In terms of real functional code that is reusable for solving common problems, we pointed at concepts such as function currying and function composition. In the next chapter, we will take a look at how functional programming can be leveraged to implement concurrent code.

10

Concurrency and Functional Programming

Concurrency is all around us, both in the real world as well as the virtual one. Humans can easily multitask (although we might not do a good job at either task). It's entirely possible to drink a cup of coffee while you are reading this chapter or to run while listening to a podcast. For machines, concurrency is a complex undertaking, although a lot of that complexity can be hidden away by the programming language we choose.

Go was built to be a language with all the necessary tools a modern-day software engineer needs. As we are now in a world where CPU power is abundant for most intents and purposes, it's only natural that concurrency was a main concern when developing the language, rather than having to bolt it on later. In this chapter, we are going to take a look at how functional programming can help with concurrency and, conversely, how concurrency can help with functional programming.

In this chapter, we are going to cover the following topics:

- Why functional programming helps us write concurrent code

- How to create concurrent functions (Filter, Map, and so on)

- How to chain functions together concurrently using the pipeline pattern

Technical requirements

For this chapter, you can use any version of Go at or above version 1.18. All the code for this chapter can be found on GitHub at https://github.com/PacktPublishing/Functional-Programming-in-Go./tree/main/Chapter10.

Functional programming and concurrency

We have already hinted at it throughout this book, but the ideas behind functional programming can help us write concurrent code. Typically, thinking about concurrency is a bit of a headache, even when a language has modern tools to support it, such as goroutines and channels. Before we dive too deep into this material, let's first take a small detour as a refresher on what exactly we mean when we talk about concurrent code, and how it compares to parallelism and distributed computing.

Concurrency, parallelism, and distributed computing

The terms *concurrency*, *parallelism*, and *distributed computing* are, at times, used interchangeably. And while they are related, they are not exactly the same thing. Let's just point out what we mean by concurrency first. **Concurrency** is what happens when our program can execute multiple tasks at the same time. For example, when we are playing a video game, typically a thread is playing audio, another one is processing input from the player, and another one is taking care of the internal game logic, updating the game state and performing the main game loop.

Video games have been around for a long time, and a game such as *DOOM* works in this way. It's also safe to say that people were not playing this on a computer with multiple cores available back in 1995. In other words, it's possible for a single core to manage the execution of these distinct tasks and give the appearance of executing them at the same time. Exactly how this is done is beyond the scope of this book, but as a takeaway, just remember that the concurrency that we will mainly focus on is concurrency as defined previously – not the simultaneous execution of code, but the concurrent execution of code. One thing to note, though, is that concurrency can happen across multiple cores, or pipelines, as well. However, to keep things simple, we can imagine concurrency using a single core.

This brings us to the second term, **parallelism**. When we talk about a program executing in parallel, this means that multiple cores are performing a task simultaneously. You can not have parallelism without a physical means to run two tasks at the same time. The native Go mechanisms of channels and goroutines are focused on concurrency and not parallelism. This is an important distinction between the two. However, Go still lends itself to building out parallel algorithms.

To get an idea of what this looks like, there are a few packages available for Go that offer parallel solutions, such as the ExaScience Pargo package: `https://github.com/ExaScience/pargo`. At the time of writing, this package is written in a pre-generics fashion, so do bear that in mind when looking through the code. In *Figure 10.1*, the difference between concurrency and parallelism is highlighted by how the tasks get executed. Notably, the two tasks in the concurrent model are broken into multiple chunks, and each gets assigned CPU time in an alternating fashion.

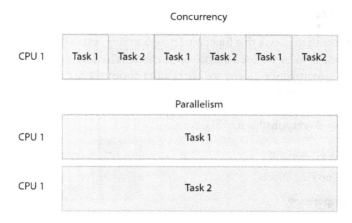

Figure 10.1: Concurrent (above) versus parallel (below) execution

Finally, we have **distributed computing**. While concurrency is part of distributed computing, it is not the only requirement for this. Distributed computation does imply spreading out computational tasks over multiple machines, in which sense it is concurrent, but there's more overhead than with typically concurrent or parallel applications.

In distributed systems, you need to have mechanisms for fault tolerance (what if one node in the network becomes unavailable?) and mechanisms for dealing with the network (unreliable or insecure networks). So, while people might talk about distributed computation as an example of concurrency, concurrency only gives you the bare minimum required. The physical infrastructure and myriad of difficulties in making a distributed system work are beyond the scope of this book. One thing to take away is that Go is a language that can be used to write distributed systems. In fact, the use of goroutines and channels might help you build out the underlying infrastructure needed for distributed systems, but you'll need more than the basic functionality of the language. If you want to learn more about distributed computing with Go, the book *Distributed Computing with Go* is a good place to start: https://www.packtpub.com/product/distributed-computing-with-go/9781787125384?_ga=2.217817046.1391922680.1675144438-1944326834.1674539572.

In this chapter, we will focus on concurrency only, and we won't zoom in on parallelism or distributed computing. However, why do we want our code to be concurrent? There are a few clear advantages that this can bring:

- **Higher responsiveness**: A program does not need to wait for a single long-running task to complete before starting another one
- **Higher performance**: If we can chunk out a heavy workload and perform this over multiple threads (and Go might schedule these across multiple cores to get a form of parallelism as well), this will reduce the time it takes to complete the operation

Functional programming and concurrency

I've made the claim before in this book that functional programming makes it easier to write concurrent code, but this claim needs to be tailored a little bit further. When talking about how functional programming makes concurrency easier, we are talking about the stricter subset of functional programming called "pure" functional programming. Pure functional programming gives us a few key features that make reasoning about concurrent execution easier and our code less error-prone. These are the main features responsible for this:

- Immutable variables and state
- Pure functions (no side effects)
- Referential transparency
- Lazy evaluation
- Composability

For the rest of this chapter, when talking about functional programming, the assumption can be made that we're talking strictly about pure functional programming. Let's focus on each of these features and explain why they make for safer concurrent code, or make our code at least easier to reason about. The result is that when our code is easier to understand, it should help us reduce the number of bugs in it.

Immutable variables and state

When working in an object-oriented model, objects typically hold an internal state. If this state is allowed to mutate, then the state that two threads are working on might diverge. By not allowing the state to change, even if operating on the same data sources (or, rather, copies of the same data), our functions can execute independently of each other without ever messing with the shared memory.

In Go, if we do want to use structs, there are some pitfalls, which we discussed in earlier chapters. By avoiding the use of pointers, we can avoid the main causes of mutation in structs. When writing pure functional code, each individual component of our code needs to be immutable. When each component is immutable, we can more safely execute functions concurrently.

Another issue we avoid by having immutable variables and states is that of resource contention. If we have a single true resource (a singleton in an object-oriented model), then this resource might be locked by thread A, causing thread B to wait until the resource is freed up before it can be used. Typically, this is implemented through a resource-locking mechanism (thread A locks the resource , X, performs operations while other threads wait for resource X, and then finally removes the lock when it is done operating). In a purely functional world, we would not need such singleton operations, partly due to our immutable state and partly due to the other benefits, such as pure functions.

Pure functions

As we saw in *Chapter 4*, a function is considered pure when it does not produce any side effects and does not interact with the outside world. In this book, we implemented many functions that are common to functional programming. All of these were written in the pure functional style (although remember that pure functional is a subset of functional programming and not strictly required). The benefits here relate to the immutable state but extend beyond it as well. If our functions do not depend on the program state, then anything modifying the state of our program cannot disrupt our function.

Beyond this, it also eliminates another class of problems. If our functions were allowed to mutate state, or our system, the order of operations would matter. For example, imagine that we were to write a concurrent function that appends content to a file. Writing to a file is a clear case of a side effect, but in a concurrent application, the content of our file would now depend on the order in which our threads are executed. This breaks the determinism of our application and, furthermore, would likely lead to a file that is not exactly what we desired. In an object-oriented model, this is again resolved through locking. In a purely functional language, the "impure" functions would be handled by monads. Go is not purely functional, but later in this chapter, we will look at the pipeline pattern through which we can model the data flow and control the side effects.

Referential transparency

Referential transparency means that we can replace a function call with its result, without changing the result of our computation. We covered this in more detail in *Chapter 2*, but for concurrency, the important aspect is that if all our calls are referentially transparent, it does not matter when exactly a call is resolved (ahead of time or just in time). This means that when we chunk our code out into concurrent functions, it is safe to resolve certain function calls ahead of time in a concurrent fashion.

Lazy evaluation

Lazy evaluation is a common approach when writing concurrent code. An example we are all too familiar with is the idea of *callbacks*. These are functions that can be passed to an asynchronous call, but they are only executed once they become relevant. It's also entirely possible for a function to never get called. For example, let's write a piece of code that performs an asynchronous GET request to a URL. We will use two callbacks, which will be lazily evaluated. The first callback will be resolved only if the GET request completed successfully, while the second callback will be resolved if the GET request failed. Note that here we mean the GET request itself did work, but we received a response code that is not in the 200 range:

```
import (
        "fmt"
        "io/ioutil"
        "net/http"
)
```

```go
type ResponseFunc func(*http.Response)
func getURL(url string, onSuccess, onFailure ResponseFunc)
    {
        resp, err := http.Get(url)
        if err != nil {
                panic(err)
        }
        if resp.StatusCode >= 200 && resp.StatusCode < 300 {
                onSuccess(resp)
        } else {
                onFailure(resp)
        }
}
```

In the preceding code, we can see that getURL requires a string representing a URL to resolve, as well as two functions. Both functions have the same ResponseFunc type, which is a function with the func(*http.Response) signature.

Next, we can write a main function in which we call getURL and provide two callbacks:

- The first callback, onSuccess, will be executed if our GET request returns a status code in the 200 range; this function will simply print out the content of the response body.

- The second callback, onFailure, will simply print an error message along with the corresponding status code that our response received. We'll call getURL twice, once with a valid URL and once with an invalid URL. However, instead of running this code synchronously, we will make the calls to getURL on separate goroutines by prefixing each call with go. This means we don't know which call will complete first, but as we are using lazy functions (a type of continuation-passing style programming), we don't have to orchestrate the control flow of our program. The correct callback will execute when its time comes. The callback, which is not necessary, will never be evaluated, so we avoid potentially expensive computation when it is not necessary:

```go
func main() {
        success := func(response *http.Response) {
                fmt.Println("success")
                content, err := ioutil.ReadAll
                    (response.Body)
                if err != nil {
                        panic(err)
                }
```

```
                  fmt.Printf("%v\n", string(content))
        }
        failure := func(response *http.Response) {
                  fmt.Printf("something went wrong,
                    received: %d\n", response
                      .StatusCode)
        }
        go getURL("https://news.ycombinator.com",
          success, failure)
        go getURL("https://news.ycombinator.com/
          ThisPageDoesNotExist", success, failure)
        done := make(chan bool)
        <-done // keep main alive
}
```

In the preceding example, our GET requests complete asynchronously and then call the corresponding callback, as defined in the getURL function. One interesting bit of code is near the end of our main snippet. We have created a bool channel, and then we are reading from this channel without ever writing to it. This essentially keeps our application alive. If we didn't have these two statements, our main function would likely exit and thus terminate our program, before our goroutines completed their computation. In a real-world application, you could also keep waiting for the threads to resolve using waitgroup. If you are stuck after running this from a terminal, press *Ctrl + C* to kill the process.

Lazy evaluation will show up again later in this chapter when we take a look at implementing functional pipes. However, we'll be looking at it more through a direct lens of concurrent applications, rather than the callback mechanisms that we saw here.

> **Threads versus goroutines**
>
> While the terms *thread* and *goroutine* are often used interchangeably, they are distinct things. Goroutines are a construct in Go, built to leverage executing tasks concurrently. They are managed by the Go runtime, are lightweight and fast to start and execute, and have a built-in communication medium (channels). Threads, on the other hand, are implemented at the hardware level and are managed by the operating system. They are slower to spin up, have no communication medium built in, and are hardware-dependent.

Composability

Functions are composable in a myriad of ways. This allows us to define the building blocks of our application and then chain them together to solve our concrete problem. As each block is independent of one another, we can build concurrency layers in between them. This will be the focus in the last part of this chapter when we will create functional pipes that can run concurrently. However, before we get there, let's take a look at making our functions internally concurrent.

Creating concurrent functions

Broadly speaking, there are two types of concurrency that we will be looking at in this chapter. We can call them **intra-concurrency** and **extra-concurrency**:

- *Intra-concurrency* is about creating functions that are implemented concurrently internal to each function. For example, in *Chapter 6*, we saw various functions such as `Filter`, `Map`, and `FMap` that lend themselves to a concurrent implementation. That will be the focus of this section. Notably, they can be used in conjunction with each other so that we achieve concurrency at multiple steps in our algorithm, and we can even decide on the level of concurrency required for each step individually.

- *Extra-concurrency* is about chaining together functions using Go's built-in concurrency features: channels and goroutines. This is explored later in the chapter.

Why are many of the fundamental building blocks of functional programming good candidates for concurrency? Well, first and foremost, it is because a purely functional implementation lends itself to a concurrent implementation without too many headaches. As we saw in the preceding chapter, concepts such as an immutable state and the elimination of side effects make it easy to take our functions and concurrently rewrite them. There should not be interference from other functions, no outside state to deal with, and no I/O to contend with. However, just because we *can* does not mean that we *should*. In this chapter, I will make the assumption that a concurrent implementation is going to be the right choice for the problems that we are solving. In the real world, concurrency is not a zero-cost implementation. There is real overhead associated with writing a concurrent application, as the threaded execution needs to be managed by our system (or, in Go's case, our runtime).

Although in Go we are not responsible for managing the goroutines ourselves, under the hood of the Go runtime, context switching is not a zero-cost implementation. This means that just adding concurrent calls does not guarantee a performance improvement and can, in fact, harm performance. Ultimately, as with anything done for performance, the key to understanding the benefit that can be achieved is obtained through profiling your application. Profiling itself is beyond the scope of this section; the only comment to make on it is that Go has built-in benchmarking tools, which we saw in earlier chapters. These can also be used to determine the cost benefit of concurrent versus sequential functions.

Concurrent filter implementation

As we started with sequential filter implementation in earlier chapters and have become more familiar with it throughout the book, let's start with this function and turn it into a concurrent implementation. Keep in mind that our initial function was a pure function, and as such, refactoring it into a concurrent one can be done without causing too much of a headache. There are a few steps to making this concurrent:

1. Split the input into batches.

2. Start a process to filter each batch.

3. Aggregate the result of each batch.

4. Return the aggregated output.

To achieve this, we do need to refactor the initial `Filter` implementation. We will leverage some of Go's built-in concurrency features to implement this, and the first thing we'll want to leverage are channels and goroutines. In our initial `Filter` function, we iterated over each element, appended it to an output slice if it matched the predicate, and finally, we returned the output slice. In this version, rather than returning an output slice, we'll write the result onto a channel:

```go
type Predicate[A any] func(A) bool
func Filter[A any](input []A, p Predicate[A], out chan []A)
    {
        output := []A{}
        for _, element := range input {
                if p(element) {
                        output = append(output, element)
                }
        }
        out <- output
}
```

Writing to a channel allows us to call this function in a traditional concurrent fashion within Go. However, before we get there, we'll have to establish a wrapper function around `Filter`, which we will call `ConcurrentFilter`. This function does a few things, including allowing us to configure the batch size. Playing around with the batch sizes can help us tweak the performance to get it where we want it (if there are too few batches, there's little benefit to running concurrently; too many, and the overhead caused by managing goroutines similarly reduces our benefit). Apart from batching our input, we'll also need to call the `Filter` function prepended with the `go` keyword so that it spins up a new goroutine. Finally, this function will read the results for each of the goroutines that we started and aggregate this result in to a single output slice:

```go
func ConcurrentFilter[A any](input []A, p Predicate[A],
    batchSize int) []A {
    output := []A{}
```

```
out := make(chan []A)
threadCount := int(math.Ceil(float64(len(input)) /
        float64(batchSize)))
fmt.Printf("goroutines: %d\n", threadCount)
for i := 0; i < threadCount; i++ {
        fmt.Println("spun up thread")
        if ((i + 1) * batchSize) < len(input) {
                go Filter(input[i*batchSize:(i+1)*batchSize],
                    p, out)
        } else {
                go Filter(input[i*batchSize:], p, out)
        }
}
for i := 0; i < threadCount; i++ {
        filtered := <-out
        fmt.Printf("got data: %v\n", filtered)
        output = append(output, filtered...)
}
close(out)
return output
}
```

In the preceding code snippet, we keep the print statements so we can see what execution looks like when running this. Let's create a simple main function that will filter a slice of integers in this fashion and look at the corresponding output:

```
func main() {
        ints := []int{1, 2, 3, 4, 5, 6, 7, 8, 9, 10}
        output := ConcurrentFilter(ints, func(i int) bool {
            return i%2 == 0 }, 3)
        fmt.Printf("%v\n", output)
}
```

Running this function gives us the following output:

```
goroutines: 4
spun up thread
```

```
spun up thread
spun up thread
spun up thread
got data: [10]
got data: [2]
got data: [4 6]
got data: [8]
[10 2 4 6 8]
```

In this output, we can see that 4 goroutines had to be spun up to process our input with a batch size of 3. This has sharded our input data into the following segments:

```
[]int{1,2,3}
[]int{4,5,6}
[]int{7,8,9}
[]int{10}
```

Next, we can see in which order the threads completed and returned their output. As you can tell from the output, we get the output back in random order. This is visible both in the got data output as well as in the final aggregated result.

> **Tip**
>
> An important callout here is that by sharding our data and running our functions concurrently, we no longer have a predictable sequence order in the output list. If we want to restore the ordering of our data, we should implement a Sort function after concurrently calling our functions.

This Filter implementation is a good template to start from when we want to make our functions run concurrently. Let's take a look at a concurrent implementation for both the Map and FMap functions.

Concurrent Map and FMap implementation

Implementing the Map and FMap functions concurrently requires the same steps as for the concurrent Filter implementation, as follows:

1. Split the input into batches.

2. Start a process to filter each batch.

3. Aggregate the result of each batch.

4. Return the aggregated output.

As such, we won't go over each step in detail for these implementations. The explanation behind each step and how we implement it is pretty much identical to the `Filter` implementation. We are showing these here for completeness and to showcase the general pattern of refactoring these functions to operate concurrently.

Concurrent Map

To implement our `Map` function concurrently, we first refactor the `Map` function that we created in *Chapter 6*. Here, again, we are removing the explicit return, and we'll use channels to communicate the output of mapping each element:

```
type MapFunc[A any] func(A) A
func Map[A any](input []A, m MapFunc[A], out chan []A) {
        output := make([]A, len(input))
        for i, element := range input {
                output[i] = m(element)
        }
        out <- output
}
```

Next, we will implement the `ConcurrentMap` function, batching the output as we did with the `ConcurrentFilter` implementation:

```
func ConcurrentMap[A any](input []A, mapFn MapFunc[A],
    batchSize int) []A {
        output := make([]A, 0, len(input))
        out := make(chan []A)
        threadCount := int(math.Ceil(float64(len(input)) /
            float64(batchSize)))
        fmt.Printf("goroutines: %d\n", threadCount)
        for i := 0; i < threadCount; i++ {
                fmt.Println("spun up thread")
                if ((i + 1) * batchSize) < len(input) {
                        go Map(input[i*batchSize:(i+1)
                            *batchSize], mapFn, out)
                } else {
                        go Map(input[i*batchSize:],
                            mapFn, out)
                }
```

```
        }
        for i := 0; i < threadCount; i++ {
                mapped := <-out
                fmt.Printf("got data: %v\n", mapped)
                output = append(output, mapped...)
        }
        close(output)
        return output
}
```

Note that both the ConcurrentFilter and ConcurrentMap implementations require batchSize to be passed as input to the function. This means that we can process each step with a different number of goroutines, and tweak each function individually:

```
func main() {
        ints := []int{1, 2, 3, 4, 5, 6, 7, 8, 9, 10}
        output := ConcurrentFilter(ints, func(i int) bool {
            return i%2 == 0 }, 3)
        fmt.Printf("%v\n", output)
        output = ConcurrentMap(output, func(i int) int {
            return i * 2 }, 2)
        fmt.Printf("%v\n", output)
}
```

In this example, we are using a batch size of 3 for filtering but only a batch size of 2 for mapping. The output of this main function looks like this:

```
goroutines: 4
spun up thread
spun up thread
spun up thread
spun up thread
got data: [10]
got data: [2]
got data: [4 6]
got data: [8]
[10 2 4 6 8]
{next statements are the output for the map function}
```

```
goroutines: 3
spun up thread
spun up thread
spun up thread
got data: [16]
got data: [20 4]
got data: [8 12]
[16 20 4 8 12]
```

Concurrent FMap implementation

This implementation is pretty similar to the Map implementation. The main difference is that our channel has changed type. Rather than having the entire function signature operate on the same A type, we'll now have a mix of A and B. This is a minor change and does not affect the implementation details beyond having to create the right type for the channels:

```
func FMap[A, B any](input []A, m func(A) B, out chan []B) {
        output := make([]B, len(input))
        for i, element := range input {
                output[i] = m(element)
        }
        out <- output
}
func ConcurrentFMap[A, B any](input []A, fMapFn ,
    batchSize int) []B {
        output := make([]B, 0, len(input))
        out := make(chan []B)
        threadCount := int(math.Ceil(float64(len(input)) /
            float64(batchSize)))
        fmt.Printf("goroutines: %d\n", threadCount)
        for i := 0; i < threadCount; i++ {
                fmt.Println("spun up thread")
                if ((i + 1) * batchSize) < len(input) {
                        go FMap(input[i*batchSize:
                            (i+1)*batchSize], fMapFn, out)
                } else {
```

```
            go FMap(input[i*batchSize:],
                fMapFn, out)
        }
    }
    for i := 0; i < threadCount; i++ {
        mapped := <-out
        fmt.Printf("got data: %v\n", mapped)
        output = append(output, mapped...)
    }
    return output
}
```

I hope that this serves as an illustration of how easy it is to create concurrent implementations for functions that are written in the purely functional style. There is one limitation posed by Go that makes this a bit more verbose than it would be in other languages. As Go is a strictly typed language (which is a good thing in general), our function signatures need to match exactly when using higher-order functions. Otherwise, we could template out the recursive part of our function and call a higher-order function for the actual implementation on each node. In pseudo-code, we would get something like the following:

```
func ConcurrentRunner(input []Input, fn func(), batchSize
    int) []Output {
    // set up channels and batch logic
    for batch in batches {
        go Run(fn(batch))
    }
    // collect output and return
}
```

Either way, we saw that leveraging concurrency in our functions is relatively headache-free and can be achieved with only a bit of refactoring. Let's move on to the final topic of this chapter, which is using concurrency mechanisms to chain functions together.

The pipeline pattern

In the previous sections, we concerned ourselves with organizing concurrency inside the functions themselves. However, we have chained them together pretty much as we would normally, by calling them in sequential order in the main function. In this section, we are going to look at the pipeline pattern, which will allow us to leverage goroutines and channels to chain function calls together. First, let's discuss what a pipeline is exactly. In 1964, Doug McIlroy wrote the following:

> *We should have some ways of coupling programs like garden hose – screw in*
> *another segment when it becomes necessary to massage data in another way.*

This quote neatly expresses the Unix philosophy of composing programs. Many of us are familiar with the concept of Unix pipes, denoted by the | symbol. By using pipes, we can chain Unix programs together. The output of one program becomes the input of the next. For example, we can use cat to read a file, and we can use wc to get the word count of that file. To join this together, we would write cat file.txt | wc. In Unix's modular program approach, the idea is that programs each serve a single purpose but can be joined together to create complex programs. This philosophy can be ported over to the functional programming paradigm. We want to chain simple functions together, where each function only has a single purpose, to create a complex program. Take the following example; each function serves a single purpose, and we chain them together using the pipe (|) character:

```
cat main.go | grep "func" | wc -l | awk '{print "lines: "
    $1}'
```

In this example, we first read the main.go file using cat. We send the content of that file to grep, which searches that content for the func keyword. Then, we send each line that matches this search to the wc program and count the lines in the output (the -l flag counts newlines). And finally, we send this to awk and print the result. What follows is a similar way of chaining Go functions together, rather than Unix commands.

Chaining functions with channels

Go ships with all the tools necessary to create such building programs, namely channels. Channels are a way to send messages (data) from one function to another; thus, we can use channels as an alternative to the Unix pipes.

The first step in creating our pipeline starts by changing how our functions get their input and output. For the rest of this chapter, we will mainly be focusing on two functions, Filter and Map, but this can be extended to any other functions. The core idea is to use channels for input and output data communication. First, let's take a look at the Filter function and how this needs to be changed to follow our channels-in/channels-out approach. We'll name our new function FilterNode. We'll get back to this naming convention later, but each function can be thought of as a node in our chain of functions. Instead of accepting a slice as input, we'll have a channel as input, from which we can read incoming data. We'll still have predicate as expected, and finally, we'll return a channel rather than a slice of data as well:

```
func FilterNode[A any](in <-chan A, predicate Predicate[A])
    <-chan A {
        out := make(chan A)
        go func() {
```

```
            for n := range in {
                    if predicate(n) {
                            out <- n
                    }
            }
            close(out)
    }()
    return out
}
```

In the preceding function, the main algorithm for filtering elements remains unchanged. We'll test each value against a predicate; if the predicate returns true, we'll keep the value (by sending it to the output channel); otherwise, we'll discard it. Pay attention to the use of the go keyword here. This function, although it gets executed immediately, is launched on its own goroutine. The function immediately returns the out channel, although the evaluation on the goroutine has not necessarily finished the computation.

The next function that we will refactor similarly is the Map function. It's an analogous change to the Filter function. We'll use a channel to receive input for the function, a channel to return the output, and run the actual mapping logic inside a goroutine, which we start before returning the channel from our function:

```
func MapNode[A any](in <-chan A, mapf MapFunc[A]) <-chan A
    {
        out := make(chan A)
        go func() {
                for n := range in {
                        out <- mapf(n)
                }
                close(out)
        }()
        return out
}
```

So far, so good – we've refactored two of our functions to fit in with this new design. Next, let's tackle the question of receiving input to these functions. From the function signature, we can tell that we need to receive data on a channel of type A. Thus, any function that can provide this can be used as the input for our function. We'll call these types of functions *generators*. The first generator that we will create takes a variadic input of type A and pushes each of these values onto a channel:

```
func Generator[A any](input ...A) <-chan A {
        out := make(chan A)
```

```
        go func() {
                for _, element := range input {
                        out <- element
                }
                close(out)
        }()
        return out
}
```

As you can tell, the main logic still resembles that of the previous `Filter` and `Map` implementations. The main difference is that we're no longer receiving values over a channel but, rather, through some other input data structure (in this case, variadic input parameters). This could also be a function that reads a file and places each line on the channel. It's similar to how `cat` worked in our earlier Unix example:

```
func Cat(filepath string) <-chan string {
        out := make(chan string)
        f, err := ioutil.ReadFile(filepath)
        if err != nil {
                panic(err)
        }
        go func() {
                lines := strings.Split(string(f), "\n")
                for _, line := range lines {
                        out <- line
                }
                close(out)
        }()
        return out
}
```

The key point is that our function places values on a channel and returns this channel. How it gets to those values doesn't matter too much for building our pipeline. Before we can test this implementation end to end, we still have one hurdle to cross. Each node in this setup writes data to a channel, but to collect the output at the end, we'll want to store it in a more common data structure. Slices are the perfect structure for this, at least in our examples. We can call this last type of function a *collector*. A collector takes a channel as input and returns a slice of the elements as output. Essentially, it's performing the reverse operation of the *generator*:

```
func Collector[A any](in <-chan A) []A {
        output := []A{}
```

```
        for n := range in {
                output = append(output, n)
        }
        return output
}
```

With this in place, we can tie all of them together into a single pipeline. To demonstrate this, in our `main` function, we will push some numbers to a channel using `Generator`. We'll then filter these numbers to only retain the even ones using `FilterNode`. These numbers then get squared using `MapNode`, and finally, we collect the output in a slice using the `Collector` function:

```
func main() {
        generated := Generator(1, 2, 3, 4)
        filtered := FilterNode(generated, func(i int) bool
            { return i%2 == 0 })
        mapped := MapNode(filtered, func(i int) int {
            return i * 2 })
        collected := Collector(mapped)
        fmt.Printf("%v\n", collected)
}
```

The output of running this is as follows:

```
[4 8]
```

The preceding is a good first step toward chaining our functions together. However, we can make it cleaner. We can build a `ChainPipes` function that will tie together the various functions and take care of managing the channels for us.

Improved function chaining

The initial approach of chaining the functions together was a workable solution, but it required some overhead, as we had to manage passing around the right channels to each subsequent function. What we want to achieve is for the engineers using our setup only needing to concern themselves with the functions to call, and which order to call them in. We don't want them to be concerned about how the channels operate underneath; we can consider that an implementation detail. What we will work toward in this section will allow us to compose the functions like so:

```
        out := pkg.ChainPipes(generated,
                pkg.CurriedFilterNode(func(i int) bool { return
    i%2 == 0 }),
```

```
        pkg.CurriedMapNode(func(i int) int { return
    i * i }))
```

This snippet gives us a bit of a teaser of what's to come. In order to chain functions like this, we will need to take advantage of function currying. Let's get there step by step. What we want to achieve is function composition by passing in functions to `ChainPipes`, as we saw in the preceding snippet. Go has a strict type system, so to make this function work nicely, we want to define a custom type for such functions, which will allow us to use it in the function signature and get the compiler to type-check for us.

The first thing we will do is define our own types for the main functions representing an operation on our data. We'll call these `Nodes`. There are three distinct types of nodes that we can define, based on the previous discussion – nodes that generate a channel, nodes that take a channel and return a new channel, and finally, nodes that take a channel and return a concrete data structure such as a slice:

```
type (
        Node[A any]          func(<-chan A) <-chan A
        GeneratorNode[A any] func() <-chan A
        CollectorNode[A any] func(<-chan A) []A
)
```

These type definitions make up the bread and butter of the function types that can be used to chain together our applications. With this in place, we can define the `ChainPipes` function as follows:

```
func ChainPipes[A any](in <-chan A, nodes ...Node[A]) []A {
        for _, node := range nodes {
                in = node(in)
        }
        return Collector(in)
}
```

The preceding snippet creates a `ChainPipes` function that takes a channel as input and a series of nodes. Finally, it will call the default collector and return the data in a slice of type `[]A`. Do note that one limitation is that we are assuming that each node has a compatible type (`A`) throughout the chain.

To make the type system work, each node needs to have the same function signature. In our initial setup, that was difficult, as we already had two distinct function signatures for `Filter` and `Map`:

```
func FilterNode[A any](in <-chan A, predicate Predicate[A])
  <-chan A
func MapNode[A any](in <-chan A, mapf MapFunc[A]) <-chan A
```

More functions would mean more distinct function signatures. Therefore, what is needed is refactoring these functions so that they adhere to the same type signature. We have already learned how to do that through function currying. We need to create two new functions that each **return** a function of type Node. Each function will have the original functionality of `Filter` and `Map` baked in but returns a new function that takes a channel as the input (hence the function is partially applied):

```
func CurriedFilterNode[A any](p Predicate[A]) Node[A] {
        return func(in <-chan A) <-chan A {
                out := make(chan A)
                go func() {
                        for n := range in {
                                if p(n) {
                                        out <- n
                                }
                        }
                        close(out)
                }()
                return out
        }
}
func CurriedMapNode[A any](mapFn MapFunc[A]) Node[A] {
        return func(in <-chan A) <-chan A {
                out := make(chan A)
                go func() {
                        for n := range in {
                                out <- mapFn(n)
                        }
                        close(out)
                }()
                return out
        }
}
```

We can tell in the preceding example that the core logic of each function has remained the same. However, rather than being instantly applied when the function is called, a new function is returned that expects to receive a channel as input and returns a channel as output. Inside this anonymous function, we have coded the `Filter` and `Map` logic respectively.

As the return type is `Node`, that means that when we call the `CurriedFilterNode` function, we are not receiving a result, but we are receiving another function that needs to be called at a later stage to actually compute the filtered list of values:

```
pkg.CurriedFilterNode(func(i int) bool { return i%2 == 0 }}
```

This is the crucial part of making our pipeline builder work. If we look at `ChainPipes` again, the main loop is calling the nodes (functions) that were supplied to it with the channel as input and reassigning the output to the same channel that was used as input:

```
for _, node := range nodes {
        in = node(in)
}
```

We could go one step further and also abstract away the generator from the `ChainPipes` function:

```
func ChainPipes[A any](gn GeneratorNode[A], nodes
    ...Node[A]) []A {
        in := gn()
        for _, node := range nodes {
                in = node(in)
        }
        return Collector(in)
}
```

With this change in place, it does imply that when calling the function, we need another curried function to supply the generator. This can be done in-line, but for clarity, the following example is a separate function existing at the package level. In this case, we will use the `Cat` function that we introduced earlier and return the curried version:

```
func CurriedCat(filepath string) func() <-chan string {
        return func() <-chan string {
                out := make(chan string)
                f, err := ioutil.ReadFile(filepath)
                if err != nil {
                        panic(err)
                }
                go func() {
                        lines := strings.Split(string(f),
                            "\n")
```

```
                    for _, line := range lines {
                            out <- line
                    }
                    close(out)
            }()
            return out
    }
}
```

Once again, this curried version of the function operates in the same way as the non-curried version. However, through currying, we can make it adhere to the type signature indicated by the ChainPipes function. We can now pass both the generator as well as the nodes to this function:

```
func main() {
        out := ChainPipes[string](CurriedCat("./main.go"),
            CurriedFilterNode(func(s string) bool {
                return strings.Contains(s, "func") }),
            CurriedMapNode(func(i string) string {
                return "line contains func: " + i }))
        fmt.Printf("%v\n", out2)
}
```

Notice that in the preceding example, we did have to give a type hint to ChainPipes to indicate the resulting type of the CurriedCat function. What we saw in the preceding section is that by using channels, the Go type system, higher-order functions, and more specifically, function currying, we can build programs by chaining together functions in the right way. Using this method of function composition, it's also easier to refactor our application. If we want to apply a map before filtering, we just need to change the order in which the node is passed to ChainPipes.

Summary

In this chapter, we took a look at how Go's concurrency model can be used when writing code in the functional paradigm. We started the chapter with a brief discussion on the difference between concurrency, parallelism, and distributed computing to delineate exactly what concurrency is.

Once we established that concurrency is the ability to do multiple tasks at once (although not necessarily simultaneously), we looked at how we can refactor the functions from *Chapter 6* into a concurrent implementation, leveraging channels and goroutines. We concluded this chapter by looking at pipelines, a way to create programs by composing functions together and orchestrating the flow of data with the use of channels. We also looked at how we can create a higher-order function to compose functions (ChainPipes) and have observed how, through the use of function currying, we can create functions that adhere to our type system without giving up type safety.

In the next and final chapter, we will take a look at programming libraries that we can leverage to create Go programs, following some of the functional programming principles that we explored in this book.

11

Functional Programming Libraries

In the previous chapters of this book, we looked at how we can leverage functional programming techniques in Go. In doing so, we have looked over how functions can be created, such as Filter, Map, Reduce, and so on. We also looked at data structures such as the monad and its application with the Maybe data type, which could represent a value that's either present or absent without having to rely on nil.

As mentioned previously, these are common tools in a functional programmer's toolbox. As such, there are open source libraries that have this functionality built in. As generics are a recent addition in Go (about 1 year ago at the time of writing), not all libraries currently leverage generics to implement these concepts. For that reason, this chapter will cover both libraries that work in all versions of Go, as well as libraries that will exclusively work in versions that support Generics.

In this chapter, we will cover the following topics:

- Pre-generics libraries for creating common FP functions
- Post-generics libraries for creating common FP functions

Technical requirements

For this chapter, any version of Go will suffice for implementing the pre-generics library code. Once we move to the post-generics libraries, a version of 1.18 or higher will be needed to support the code. All the code can be found on GitHub at https://github.com/PacktPublishing/ Functional-Programming-in-Go./tree/main/Chapter11.

There are a few things to call out before we dive deeper into this topic that relate somewhat to the technical requirements.

Is the library alive – and do the examples still match it?

When writing a book about a specific programming language, it is hard to write it in an evergreen fashion. But programming libraries are perhaps even harder to keep evergreen than any other content. There are two reasons for this, which are important to acknowledge:

- The implementations can change, and versioning is not always respected.

- The library may become unsupported in the future.

The first problem, *changing implementations*, should be somewhat mitigated by the fact that only popular libraries will be explored in this chapter, whereby popularity is judged by engagement on GitHub as well as stars on GitHub. It's an imperfect measure, but it's better than not having anything to go by.

I hope that these libraries respect versioning and that they limit breaking changes as much as possible. Still, I can't guarantee that these libraries won't change and that the functions will work as-is when you are reading this chapter. In the code example, I will highlight which version of the library is being shown so that the results can at least be recreated by fetching the correct version of the library, even if that is not the latest version. This brings us to the second, *related* problem.

The library may become unsupported. If you are working with an older version of the library to recreate the examples in this chapter because the latest version introduced some breaking changes, then clearly there's a risk that you'll run into some known issues, and you might not get support as you're using an older version. But, even if the examples shown here work correctly with the latest version of the library, the library can still be stale. If everything works as intended and the library is considered feature complete, that's not immediately a red flag.

However, it does mean that finding these libraries can be hard. The best way to determine this is by looking for any activity on the GitHub (or GitLab) page. For example, is the most recent commit only a few days or weeks ago, or is it years in the past? Are the contributors actively responding to tickets or do they all go unanswered? Are they engaging with their community over Discord or IRC? These are all examples that can hint at how well maintained a library is.

Legal requirements

I'll keep this part brief, as I am not a lawyer. But anyone dealing with open source code should be aware that not all open source code is permissive.

> **Note**
> Before working with a library, especially in a commercial context, be sure to review the software license and confirm that your use case is legally allowed and under which conditions. (For example, some licenses will allow the use of the code with attribution. Others will only allow for non-commercial use cases and so on.)

Pre-generics libraries for creating common FP functions

With or without generics, it is common to operate on collection-style data structures in any programming language. Storing a series of values, whether it is a list of numbers representing scores on a test or a collection of structs such as all employees working in a hospital, is common enough that you'll run into these data structures sooner rather than later. The operations that are performed on these can also fall into a few categories, especially once we abstract them into higher-order functions. You either have to modify the data elements in some way (for example, multiplying all the values by two) or modify the container in some way (for example, removing all the odd numbers). As we have seen, rather than implementing a function such as `removeOdds` or `multiplyNumbers`, what we'd like to write is just a function that can filter any element based on a predicate or change an element based on a transformation (these are the Filter and Map functions, respectively).

Before generics were introduced, there was no clear and best way to handle this. The reasoning for not abstracting these use cases, at the time, was that writing functions specific to your data structure would deliver the best results in terms of performance. So, you'd give up a bit of developer comfort but would get a more performant application in return. In hindsight, many of the operations on collections have an identical implementation, which means there's no real performance difference. It's only a natural consequence that people came up with ways of building abstractions for repeating implementations.

Broadly speaking, there are two ways this problem could be tackled before the introduction of generics – either by programming against the empty interface (`interface{}`), an interface that any data type implicitly adheres to in Go, or through code generation. The former, programming against `interface{}`, has too many disadvantages in terms of typesafety and runtimesafety to advocate strongly for it. But the latter, code generation, is still interesting to look at, if only because code generation could still be useful in a post-generics world albeit for different use cases.

> **Libraries versus custom implementations**
>
> In this book, we have seen ways to create our own set of functions that follow the functional programming paradigm. Libraries might offer a more efficient implementation and can prevent you from reinventing the wheel. However, if you want to keep your dependency graph lightweight, it is much easier to do so by providing a few implementations yourself now that Go has: generics. In a pre-generics version of Go, this was much harder and I'd favor the library-based approach. Neither the empty interface-based approach nor the code generation approach is easy to implement without errors and headaches.

Code generation libraries for pre-generics Go

Code generation, as the name implies, is a technique for generating Go code that can then be used like regular Go code in our application. The Go toolchain has all the necessary tools to do this out of the box. In Go, it is possible to add comments to your code that the compiler will interpret as commands. Such comments make it possible to trigger a special operation during the compile time of your program. These comments are called **pragma**. For example, you can add a comment to a function that will tell the compiler to avoid in-lining this function (the compiler can ignore it, so it's more a suggestion than a command):

```
//go:noinline
func someFunc() {}
```

The idea behind the code generation library, which we will explore shortly, is that using these special comments can trigger the generation of functions for a specific type, which implements the common functional programming operations such as filtering, mapping, reducing, and so on. The first library that we will explore, Pie, works in exactly this way.

A slice of Pie

The library that we will explore is **Pie**, written by Elliot Chance and available on GitHub here: https://github.com/elliotchance/pie/tree/master/v1. This library is available in two versions:

- Version 1 focuses on Go at or below 1.17
- Version 2 is the newer version for working with generics and needs Go 1.18 or above to work

In version 1, there are two ways of using this library. You can either use the functions directly to operate on common data types ([]string, []float64, or []int), or you can use this library to generate functions for your own data type. First, we'll explore the built-in structures and then move on to generating functions for custom types.

Using the built-in functions of Pie

Pie supports built-in functions for three data types:

- []string
- []float64
- []int

These are quite common, so it makes sense that these are supported by default. In the examples throughout this book, we have shown how we can filter a slice of integers to retain only the even numbers. Then, we squared them by using the Map function. Doing this in Pie is easy and follows the same idea as the code that we implemented in *Chapter 6* and beyond. As we are doing this through the use of a library, let's first take a look at the content of the go.mod file to highlight which version of Pie we are using:

```
go 1.17
require github.com/elliotchance/pie v1.39.0
```

> **Note**
> This is showing `go 1.17` as we are explicitly looking at libraries that can be used before generics were introduced.

Now that we have imported the library (after running `go get`), we can use it in our application. Let's build the Filter and Map example, as explained earlier:

```go
package main
import (
        "fmt"
        "github.com/elliotchance/pie/pie"
)
func main() {
        out := pie.Ints{1, 2, 3, 4, 5, 6, 7, 8, 9, 10}.
                Filter(func(i int) bool {
                        return i%2 == 0
                }).
                Map(func(i int) int { return i * i })
        fmt.Printf("result: %v\n", out)
}
```

Running this code will output `result: [4 16 36 64 100]`, as expected. Pie allows us to build and chain functions together, similar to what we have seen in this book so far. Out of the box, this only works for slices of strings, ints, and float64s. Each of these requires a custom implementation in the library. By attaching the function to a concrete type, it can support multiple Filter and Map functions defined for distinct data types. This is also something that we have looked at doing ourselves, and as pointed out, this is a time-consuming and repetitive undertaking.

What Pie does is remove some of this repetitive work by using code generation to generate the implementation for each data type. The details of how code generation works in this library are beyond the scope of this book, but I'd encourage checking out the library itself on GitHub and diving into the code to get a better appreciation for how this was built as it is genuinely quite interesting.

Pie ships with a lot of functions. To get an up-to-date listing with a description of each, take a look at the wiki at `https://github.com/elliotchance/pie/tree/master/v1`.

Pie for custom data types

If we want to use Pie for our own data type, we need to generate the code to do this:

1. First, let's set up a struct that we can use in all the following examples. We'll create a struct to represent a dog, and also a type alias for a slice of the [ ] Dog type:

    ```go
    //go:generate pie Dogs.*
    type Dogs []Dog

    type Dog struct {
        Name string
        Age  int
    }
    ```

2. With this set up, we can run the go generate command and generate all of Pie's functions for our custom data type. This created a new file, dogs_pie.go, in the same directory as our type definitions. By looking through the generated file, we can see which functions were generated. For example, the Reverse function was generated specifically for the Dog data type. This is copied verbatim here:

    ```go
    // Reverse returns a new copy of the slice with the
    //   elements ordered in reverse.
    // This is useful when combined with Sort to get a
    //   descending sort order:
    //
    //    ss.Sort().Reverse()
    //
    func (ss Dogs) Reverse() Dogs {
    // Avoid the allocation. If there is one element or
    //    less it is already
    // reversed.
    if len(ss) < 2 {
            return ss
    }
    sorted := make([]Dog, len(ss))
    for i := 0; i < len(ss); i++ {
        sorted[i] = ss[len(ss)-i-1]
    }
    return sorted
    }
    ```

3. We can also find the Filter and Map functions defined for the Dog data type. Again, these have been copied verbatim but with the comments omitted:

```
func (ss Dogs) Filter(condition func(Dog) bool) (ss2
     Dogs) {
  for _, s := range ss {
     if condition(s) {
  ss2 = append(ss2, s)
       }
  }
  return
}


func (ss Dogs) Map(fn func(Dog) Dog) (ss2 Dogs) {
  if ss == nil {
     return nil
  }
  ss2 = make([]Dog, len(ss))
  for i, s := range ss {
     ss2[i] = fn(s)
  }
  return
}
```

What this approach should highlight is that if you have many distinct types for which you are generating these functions, you are polluting your code base quite a bit with similar but not quite identical code. The executables that you are building will be larger as a result of this, and while it's not often something that you have to think about anymore, if you are targeting a platform with limited memory availability, this might be a showstopper.

That said, let's take a look at how we can use the generated functions with another example in the main function. First, we'll create some dogs, each with a name and an age. Then, we will filter the dogs for those that are older than 10. These results will then get sorted based on age, and this will be printed as the result:

```
func main() {
        MyDogs := []pkg.Dog{
                pkg.Dog{
                        "Bucky",
```

```
                            1,
                    },
                    pkg.Dog{
                            "Keeno",
                            15,
                    },
                    pkg.Dog{
                            "Tala",
                            16,
                    },
                    pkg.Dog{
                            "Amigo",
                            7,
                    },
            }
        results := pkg.Dogs(MyDogs).
                Filter(func(d pkg.Dog) bool {
                        return d.Age > 10
                }).SortUsing(func(a, b pkg.Dog) bool {
                return a.Age < b.Age
        })
        fmt.Printf("results: %v\n", results)
}
```

Given this input, we get the following output:

```
results: [{Keeno 15} {Tala 16}]
```

There are more functions to explore in Pie for a pre-generics version of Go. But let's shift our focus now to contemporary Go code and look at libraries that we can leverage since Go 1.18.

> **go generate and go environment**
>
> To run go generate with Pie or any other executable that you download through go get, you need to ensure that your environment setup has been configured correctly to discover such executables. In a *nix-based system, this means that go/bin needs to be added to the $PATH variable. On Windows, you need to add go/bin to the environment variables. In the worst case, you can either download the GitHub source code or look for the directory where go dependencies are downloaded and build them yourself through go install and then move the executable to an environment location that is registered for your system.

> **Pie and Hasgo**
>
> For the sake of being transparent, there is another library out there that follows a similar approach to Pie but tailors the functions to a Haskell-like implementation. This library is called **Hasgo** (`https://github.com/DylanMeeus/hasgo`), of which I am the author. While both libraries work similarly, Pie offers more functions out of the box and fully supports Go 1.18. But if you have written Haskell before, Hasgo might feel more familiar in terms of function naming and documentation.

Post-generics functional programming libraries

Functional programming libraries have seen a rise in popularity since the advent of generics in Go. No longer is it necessary to mess with the empty interface or to rely on code generation to build out the staples that make up functional programming languages. We'll explore a few libraries in this section and see how their implementation compares. In doing so, we will stick with examples that are more or less identical but might show off some different functions from the ones we have seen so far in this book.

Pie with generics

The first library that we will look at is Pie. In the previous section, we indicated that there are two versions of Pie available today: v1, which is tailored to Go before the introduction of generics, and v2, which offers the same functionality in terms of functions but leverages generics to do so. v2 is actively maintained, so I expect that over time v1 and v2 will no longer offer feature parity. That said, the Go community is pretty good at adopting the latest Go version wherever possible, so I don't expect this to be a blocker for anyone.

Before we dive into the code, this is a snippet of the go.mod file, just to highlight which version of Pie we are using:

```
go 1.18
require github.com/elliotchance/pie/v2 v2.3.0
```

The go 1.18 statement indicates that we can use generics, as generics were introduced in this version. Any version above 1.18 will work for the examples that we are about to see.

As with the pre-generics example, we will work with the Dog struct and a slice of the [] Dog type. Unlike the previous non-generics example, we don't need to add the compiler pragma to generate any code, nor do we need the type alias for [] Dog (although using this can still be good practice in a real application):

```
type Dog struct {
        Name string
        Age  int
}
```

In the main function, we will create a slice of dogs. Then, we will once again filter for the dogs that are older than 10. We will then map their name to uppercase and finally return the result sorted by age:

```
import "github.com/elliotchance/pie/v2"
func main() {
        MyDogs := []Dog{
                Dog{
                        "Bucky",
                        1,
                },
                Dog{
                        "Keeno",
                        15,
                },
                Dog{
                        "Tala",
                        16,
                },
                Dog{
                        "Amigo",
                        7,
                },
        }
        result := pie.Of(MyDogs).
                Filter(func(d Dog) bool {
                        return d.Age > 10
                }).Map(func(d Dog) Dog {
                d.Name = strings.ToUpper(d.Name)
                return d
        }).
                SortUsing(func(a, b Dog) bool {
                        return a.Age < b.Age
                })
        fmt.Printf("out: %v\n", result)
}
```

As you can tell, the code is pretty similar to the pre-generics version. However, no code generation was used to achieve this. Also, note that `pie.Of()` figured out what type of data we are operating on. In the pre-generics version, this is part of the reason why we had to create a type alias for `[]Dog` – so that the code generator could then use Filter, Map, Reduce, or some other method for the correct slice type and attach it for dot notation-style function chaining. With generics, we no longer need to do so. In general, Pie is a good library to explore if you want to introduce generics to a team, as the familiar dot notation-style chaining of function calls looks natural to developers who are used to an object-oriented approach. As mentioned previously, it has an extensive set of functions that can be used out of the box. Next, let's look at a library for functional programming that is based on **Lodash**.

Lodash, for Go

lo (`https://github.com/samber/lo`) is a library that, similar to Pie, adds easy-to-use functions to Go and is quite popular at the moment. It is inspired by the insanely popular Lodash library for JavaScript (`https://github.com/lodash/lodash`), which currently has over 55,000 stars on GitHub and is widely used.

Currently, lo supports 38 functions that operate on slices, 16 of which operate on the Map data type, and a bunch of convenience functions for searching, tuples, channels, and (set) intersection-style operations. It's not practical to outline all the functions here, but if you have a problem that requires operating on these common container data types, it's a good idea to check whether this library suits your needs before reinventing the wheel. What we will do in this section is take a look at a similar example to the one we used for Pie.

An example implementation with lo

As we are importing a new library, the following snippet shows the library and version that we will use for these examples:

```
go 1.18
require (
        github.com/samber/lo v1.37.0
)
```

To demonstrate this library, we'll once again use a `main` function and a slice of dogs. In this case, we'll want to do the following. First, we'll deduplicate the slice so that each element in the slice is unique. Then, we will transform the names of all dogs into uppercase variants. This is the result that we will print:

```
func main() {
        result :=
                lo.Map(lo.Uniq(MyDogs), func(d Dog, i int)
                Dog {
```

```
                        d.Name = strings.ToUpper(d.Name)
                        return d
            })
        fmt.Printf("%v\n", result)
  }
```

In this small example, you can see how the use of the library is more reminiscent of a style chosen by (pure) functional programming languages rather than the dot notation style common to object-oriented code. We are chaining function calls by passing them as the input parameters of the higher-order function. Note that these are not lazily evaluated. In the preceding example, first, the `Uniq` function runs, which removes the duplicate entries from our input slice. Then, the `Map` function runs and applies the transformation. Remember that we are mutating the **copy** of the `Dog` struct by calling `d.Name = ...`, but this does not mutate the original data element. We explored this in more detail in previous chapters of this book.

There is one additional feature that bears calling out. `lo` contains a subset of the library supported for concurrent function calls. There is a package in `lo` under `lo/parallel` that supports the parallel evaluation of function calls. Let's rewrite our example but have the `Map` function work concurrently. (**Also, note that this package is called parallel but is talking about concurrent code**).

First, here's the import statement and import alias:

```
        lop "github.com/samber/lo/parallel"
```

Next, here's the code to run the `Map` function concurrently, with the `Uniq` function still running sequentially:

```
        result :=
            lop.Map(lo.Uniq(MyDogs), func(d Dog, i int)
                Dog {
                    d.Name = strings.ToUpper(d.Name)
                    return d
            })

        fmt.Printf("%v\n", result)
```

This took almost no refactoring from our side but leveraged goroutines for concurrency. Pretty neat!

To close this chapter, let's look at a library by the same author of `lo` that contains monad-like data structures such as the `Maybe` data type, which we explored in *Chapter 5*.

Mo, for go

Mo is a library that adds support for monad-like data structures in Go and is relatively popular. It fully supports Go 1.18+, and thus is built around generics. You can find the package itself here: `https://github.com/samber/mo`.

It's worth taking the time to explore this library and read the documentation, especially as this could have changed by the time you read this book. In essence, it works in the same way as the `Maybe` implementation in *Chapter 5* although, in this library, that type is called `Option`. We can create a data type that optionally contains a value, but can also represent the absence of a value. This data type then supports functions to transform the data or get the data in a nil-safe way. For example, let's create an option that contains a dog:

```
func main() {
        maybe := mo.Some(Dog{"Bucky", 1})
        getOrElse := maybe.OrElse(Dog{})
        fmt.Println(getOrElse)
}
```

This prints the following:

```
{Bucky 1}
```

Now, if we were to use this to represent a `nil` value, we could still access it in a type-safe way. The `OrElse` function will ensure that a backup is used as a result of the function call, which is the default value provided by the caller. For example, let's say we add the following code to our `main` function:

```
        maybe2 := mo.None[Dog]()
        getOrElse2 := maybe2.OrElse(Dog{"Default", -1})
        fmt.Println(getOrElse2)
```

The output would look like this:

```
{Default -1}
```

This library supports other types as well, such as `Future` and `Task`. But one particularly useful one is the `Result` type, which is more or less like the `Maybe` type but is meant to work in cases where a value can optionally contain an error. We'll demonstrate this in the following snippet. First, we will call the `Ok()` function, which creates the `Result` type with a valid `Dog` object. In the second case, we will create the `Result` type with an error instead of a `Dog` object. In both cases, we will try to get and print the result, as well as the error message:

```
        ok := mo.Ok(MyDogs[0])
        result1 := ok.OrElse(Dog{})
```

```
err1 := ok.Error()
fmt.Println(result1, err1)
err := errors.New("dog not found")
ok2 := mo.Err[Dog](err)
result2 := ok2.OrElse(Dog{"Default", -1})
err2 := ok2.Error()
fmt.Println(result2, err2)
```

If we run this function, we will get the following output:

```
{Bucky 1} <nil>
{Default -1} dog not found
```

This shows us that based on the content of the `error` value for `Result`, the behavior of the type is different. In the first instance, where we don't have an error, we get back the correct dog and the error is empty. In the second instance, we get back the default value that we provided as part of the `OrElse` statement, as well as the underlying error message.

Summary

In this chapter, we looked at libraries that implement concepts of the functional programming paradigm. We started by looking at Pie, a library that can help users in building code in the functional paradigm whether working with a code base that uses Go before or after the introduction of generics in Go 1.18. Specifically for the pre-generics version, we looked at the approach of code generation for custom types to get generics-like behavior. Pie allowed us to showcase the ease with which we can create functions such as Map and Filter since the introduction of generics.

Then, looked at the Lodash-inspired Go library, `lo`. This library supports common functions that operate on container data types such as slices and maps, but unlike Pie, it follows a nested approach to function chaining rather than the dot notation syntax. `lo` does offer concurrent implementations for certain functions, so if performance is a concern and concurrency seems like the right solution, checking out this library is a good idea.

Finally, we looked at the `mo` library, which adds monad-like data structures to Go. Specifically, we looked at the `Option` data structure, which is comparable to the `Maybe` data structure that we created in *Chapter 5*. `mo` also offers a `Result` type, which is built for error handling and allows us to program more safely when dealing with potential `error` values.

Index

www.packtpub.com

Subscribe to our online digital library for full access to over 7,000 books and videos, as well as industry leading tools to help you plan your personal development and advance your career. For more information, please visit our website.

Why subscribe?

- Spend less time learning and more time coding with practical eBooks and Videos from over 4,000 industry professionals

- Improve your learning with Skill Plans built especially for you

- Get a free eBook or video every month

- Fully searchable for easy access to vital information

- Copy and paste, print, and bookmark content

Did you know that Packt offers eBook versions of every book published, with PDF and ePub files available? You can upgrade to the eBook version at www.packtpub.com and as a print book customer, you are entitled to a discount on the eBook copy. Get in touch with us at customercare@packtpub.com for more details.

At www.packtpub.com, you can also read a collection of free technical articles, sign up for a range of free newsletters, and receive exclusive discounts and offers on Packt books and eBooks.

Other Books You May Enjoy

If you enjoyed this book, you may be interested in these other books by Packt:

Domain-Driven Design with Golang

Matthew Boyle

ISBN: 9781804613450

- Get to grips with domains and the evolution of Domain-driven design
- Work with stakeholders to manage complex business needs
- Gain a clear understanding of bounded context, services, and value objects
- Get up and running with aggregates, factories, repositories, and services
- Find out how to apply DDD to monolithic applications and microservices
- Discover how to implement DDD patterns on distributed systems
- Understand how Test-driven development and Behavior-driven development can work with DDD

Event-Driven Architecture in Golang

Michael Stack

ISBN: 9781803238012

- Understand different event-driven patterns and best practices
- Plan and design your software architecture with ease
- Track changes and updates effectively using event sourcing
- Test and deploy your sample software application with ease
- Monitor and improve the performance of your software architecture

Packt is searching for authors like you

If you're interested in becoming an author for Packt, please visit authors.packtpub.com and apply today. We have worked with thousands of developers and tech professionals, just like you, to help them share their insight with the global tech community. You can make a general application, apply for a specific hot topic that we are recruiting an author for, or submit your own idea.

Share your thoughts

Now you've finished *Functional Programming in Go*, we'd love to hear your thoughts! Scan the QR code below to go straight to the Amazon review page for this book and share your feedback or leave a review on the site that you purchased it from.

https://packt.link/r/9781803238012

Your review is important to us and the tech community and will help us make sure we're delivering excellent quality content.

Download a free PDF copy of this book

Thanks for purchasing this book!

Do you like to read on the go but are unable to carry your print books everywhere?

Is your eBook purchase not compatible with the device of your choice?

Don't worry, now with every Packt book you get a DRM-free PDF version of that book at no cost.

Read anywhere, any place, on any device. Search, copy, and paste code from your favorite technical books directly into your application.

The perks don't stop there, you can get exclusive access to discounts, newsletters, and great free content in your inbox daily

Follow these simple steps to get the benefits:

1. Scan the QR code or visit the link below

https://packt.link/free-ebook/9781801811163

2. Submit your proof of purchase
3. That's it! We'll send your free PDF and other benefits to your email directly

www.ingramcontent.com/pod-product-compliance
Lightning Source LLC
Chambersburg PA
CBHW060542060326
40690CB00017B/3579